AN INTRODUCTION TO COMPLEX ANALYSIS

An Introduction to Complex Analysis

Peter L. Walker
Department of Mathematics
University of Lancaster

A HALSTED PRESS BOOK

JOHN WILEY & SONS
New York

First Published October 1974

Published in the USA by
Halsted Press, a Division of
John Wiley & Sons, Inc.
New York

Library of Congress Cataloging in Publication Data

Walker, Peter L. 1928–
An introduction to complex analysis.
"A Halsted Press book."
Bibliography: p.
Includes index.
1. Mathematical analysis. 2. Functions of complex variables. I. Title.
QA300.W29 1975 515′.9 74–24686
ISBN 0–470–91807–1

Published in Great Britain by
Adam Hilger Ltd
Rank Precision Industries
29 King Street, London WC2E 8JH
Printed in Great Britain

Contents

Preface

The elementary part of complex analysis may be described as those concepts which are necessary to comprehend, prove, and apply the residue theorem. As such it comprises Cauchy's theorem and the integral formulae for regular functions, power series developments, classification of singularities, the topological index or winding number and the residue theorem together with its consequent theorems such as that of Rouché.

The aim of this book is to give an account of these topics which is both concise and mathematically complete—that is, the results are both rigorously established and independent of those topological difficulties which are discussed in an appendix.

The first chapter contains elementary properties of open and closed sets in the plane, and also the definition and properties of the basic analytic operations of integration and differentiation as applied to complex functions. Any integration process which can handle piecewise continuous functions is adequate here: Appendix A indicates one possibility.

Chapter 2 contains a proof of Cauchy's theorem in a form adequate for our later applications. The hypothesis of 'starredness' serves as a beginner's version of 'simple connectedness', and allows a proof which is purely analytical in character.

There follow two short chapters which contain the derivations of the classical power series expansions of Taylor and Laurent, and in which these are used to classify isolated singularities and define the notion of a residue.

The culmination of this development come in Chapter 5. Here the properties of the topological index are established, including a reasonably general criterion for its evaluation. The residue theorem is then proved, applied to the evaluation of integrals and series, and used to derive Rouché's theorem and its corollaries.

The final chapter gives an introduction to harmonic functions and the Dirichlet problem which is fundamental to the theory of potential in classical applied mathematics. This chapter ends with a proof of a precise version of the Riesz conjugation theorem which to the best of the author's knowledge has not appeared in book form before.

The three appendices cover respectively an elementary description of the so-called regulated integral, a discussion of some of the topological problems encountered in proving our results in greater generality (the hypothesis that a curve should be simple, or non-self-intersecting is considered here for the first time) and the complex logarithmic and argument functions.

As a prerequisite for reading this book, the reader should have a working knowledge of the elementary facts of analysis relating to functions of a single real variable, the convergence of sequences and series and the properties of uniform convergence including the Weierstrass M-test.

The material of Chapters 2 to 5, preceded by a suitable selection from Chapter 1, has been given in various forms to undergraduates in the second year of their honours degree at Lancaster University, and is designed for a course of perhaps twenty-five lectures at this level. Many succeeding classes of students have helped to modify and improve the form in which the material is here presented.

It is a great pleasure to acknowledge both the detailed comments on an early draft which were made by a former colleague, Dr. (now Professor) I. J. Maddox, and the skill and efficiency of Mrs. S. Brennan and Mrs. J. Howard, who typed the manuscript.

Lancaster, 1974. Peter L. Walker

1

Basic Properties of Sets and Functions in the Complex Plane

The first part of this chapter contains the elementary properties of subsets of the complex plane, such as openness, compactness, or connectedness, which arise because of the metric structure of the plane. No formal knowledge of metric space theory is assumed, although many of the arguments are equally valid in the more general context. This part should probably be gone over lightly (if at all) by most readers. The second part of the chapter treats the definitions of the analytic processes of integration and differentiation applied to complex functions. These are dealt with in some detail, since the distinction between the real and complex cases is often crucial.

§1. METRIC PROPERTIES OF THE COMPLEX PLANE

We use the notations **Z**, **R** and **C** for the sets of integers (positive and negative), real numbers and complex numbers. A complex number z uniquely determines a pair (x, y) of real numbers, called its real and imaginary parts respectively, and in accordance with normal usage we write $z = x + iy$. This correspondance allows us to identify **C** with $\mathbf{R}^2$, two-dimensional real Euclidean space, in a natural way.

We assume that the reader will be familiar with the usual operations of addition and multiplication for complex numbers, and with the fact that they form a field under these operations. It is to be emphasized that the ordering of the real line does *not* carry over to the complex plane. (If i were either >0 or <0, we should have to have $i^2 > 0$, or $-1 > 0$, a contradiction.) A statement such as '$r > 0$' will therefore automatically read 'r is real and positive'.

We also assume throughout that the reader is familiar with the usual set theoretic operations of union, intersection, set theoretic differences, and the idea of defining a set by prescribing properties of its elements: for example

$$\{z : z = x + iy, x \geqslant 0, y \geqslant 0, x^2 + y^2 \leqslant 1\}$$

describes those points which lie in the first quadrant of the unit disc in the complex plane.

The complex conjugate $\bar{z}$ of z is defined as $\bar{z} = x - iy$, so that the real and imaginary parts, denoted $\operatorname{Re} z$ and $\operatorname{Im} z$ respectively, are given by

$$x = \operatorname{Re} z = \tfrac{1}{2}(z + \bar{z}),$$

$$y = \operatorname{Im} z = \frac{1}{2i}(z - \bar{z}).$$

The modulus $|z|$ is defined as the positive square root of $z\bar{z} = x^2 + y^2$, and has the following properties:

(i) $|z| \geqslant 0$, and $|z| = 0$ if and only if $z = 0$,
(ii) $|z_1 z_2| = |z_1||z_2|$,
(iii) (the triangle inequality) $|z_1 + z_2| \leqslant |z_1| + |z_2|$, with equality if and only if either z_1 or z_2 is zero, or the ratio z_1/z_2 is real and positive (this last assertion has an obvious geometrical significance—as does the inequality itself),
(iv) if $z = x + iy$, then

$$\max(|x|, |y|) \leqslant |z| \leqslant |x| + |y| \leqslant 2\max(|x|, |y|).$$

The modulus allows us to define the distance between complex numbers z_1 and z_2 as $|z_1 - z_2|$. If $z_1 = x_1 + iy_1$, $z_2 = x_2 + iy_2$, then

$$|z_1 - z_2| = \sqrt{\{(x_1 - x_2)^2 + (y_1 - y_2)^2\}},$$

so that the distance coincides with the familiar Pythagorean distance in $\mathbf{R}^2$. We can also define sphere and circles in the complex plane in the following way.

Definition 1.1. Let a be a complex number, and $r > 0$. (As we mentioned earlier, this means that r is a positive *real* number.)

The open sphere, $S(a, r)$, is defined by

$$S(a, r) = \{z : |z - a| < r\},$$

that is $S(a, r)$ is the set of all points whose distance from a is less than r. Similarly the closed sphere

$$\bar{S}(a, r) = \{z : |z - a| \leqslant r\};$$

and the circle, or circumference

$$C(a, r) = \{z : |z - a| = r\}.$$

These definitions evidently bear out our intuitive ideas of what (two dimensional) spheres and circles should be. We shall call a the centre of the sphere, and r its radius. We use the open spheres to define properties of more general sets in the next definition.

Definition 1.2. (i) A set $G \subset \mathbf{C}$ is called open, if for each point a in G, there is an open sphere centred at a, contained in G. To show that an open sphere is itself an open set in accordance with this definition (and hence that the terminology involves no inconsistency), let b be any point of $S(a, r)$. Then by definition $|b - a| < r$, and so if $r_1 = r - |b - a|$, then r_1 is real and >0. The triangle inequality now shows that $S(b, r_1) \subset S(a, r)$ and hence that $S(a, r)$ is an open set in accordance with the definition. It is worth noting that both $\varnothing$ (the empty set) and $\mathbf{C}$ itself are open sets. We say that a set G is a neighbourhood of a point a if G is open and $a \in G$.

(ii) We say that a sequence (a_n) of points in the complex plane converges to z_0 as $n \to \infty$, if $|z_n - z_0| \to 0$ as $n \to \infty$.

If E is any set in $\mathbf{C}$, and z_0 a point of $\mathbf{C}$ which may or may not be in E, we say z_0 is a limit point of E if there is a sequence (z_n) of points of E distinct from z_0, with $z_n \to z_0$ as $n \to \infty$.

It is an easy exercise in the definitions to show that z_0 is a limit point of E if and only if for each $r > 0$, $S(z_0, r)$ contains points of E other than z_0.

(iii) A set $F \subset \mathbf{C}$ is called closed if every limit point of F is an element of F. Again it is an easy exercise in definitions (i) and (ii) to show that a set F is closed if and only if its complement $\mathbf{C} \setminus F$ is open. It is worth emphasizing that this condition is not the same as saying 'F is not open', for sets such as $\{z: 0 < |z| \leqslant 1\}$ are neither open nor closed.

A similar argument to that under (i) shows that for any $a \in \mathbf{C}$ and $r > 0$, the set $\{z: |z - a| > r\}$ is open, and hence its complement, $\bar{S}(a, R)$ is closed. The use of the description 'closed sphere' for $\bar{S}(a, R)$ is therefore consistent with our definition of a general closed set.

(iv) A set $B \subset \mathbf{C}$ is said to be bounded if for some $R > 0$, $B \subset S(0, R)$.

(v) A set $K \subset \mathbf{C}$ is said to be compact if from every collection of open sets which contains K in its union (we say the collection 'covers' K), we can pick out a finite number whose union also covers K (we say we can find a 'finite subcover').

Since compactness is a property of sets which is not easily recognized from the definition, it is important to have an alternative description of a compact set. This is supplied by the following well-known result.

Theorem (Heine–Borel). *A set $K \subset \mathbf{C}$ (or more generally in $\mathbf{R}^n$, $n = 1, 2, 3, \ldots$) is compact if and only if it is closed and bounded.*

Unlike our assertions in (ii) and (iii), this result is far from immediate. We show that the conditions stated are necessary, and refer the reader to the book by Rudin mentioned in the bibliography for the sufficiency.

Suppose then that K is compact. The whole space $\mathbf{C}$ is the union of the open spheres $S(0, k)$ for $k = 1, 2, \ldots$, and hence these spheres form an open cover of K. From the definition of compactness we can find a finite subcover for K: if k_1 is the largest radius of a sphere in the subcover, $K \subset S(0, k_1)$ and so is bounded.

To show K is closed, we use a similar argument. Suppose a point a is in the complement of K. Then as we observed under (iii) the sets

$$\left\{z: |z - a| > \frac{1}{n}\right\}$$

are open, and

$$\mathbf{C} \setminus \{a\} = \bigcup_{n=1}^{\infty} \left\{z: |z - a| > \frac{1}{n}\right\}.$$

Since $a \notin K$, these sets form an open cover of K, and hence since K

is compact there is a finite subcover. If $1/k_2$ is the smallest radius in the subcover, then

$$K \subset \left\{z: |z - a| > \frac{1}{k_2}\right\} \quad \text{or} \quad S\left(a, \frac{1}{k_2}\right) \cap K = \varnothing.$$

Hence the complement of K is open, and so K is closed.

The equivalence of compactness with closure and boundedness has a number of important consequences, which we shall need later.

(A) (Bolzano–Weierstrass property). *Let E be a bounded infinite set in* **C**. *Then E has a limit point in* **C**.

We establish this statement in its contrapositive form, namely that a bounded set E with no limit point must be finite. Suppose then that E is such a set. Since it is bounded, $E \subset \bar{S}(0, R)$ for some value of R, and $\bar{S}(0, R)$ is closed and bounded, hence compact by the Heine–Borel theorem. Given any point z of $\bar{S}(0, R)$, our hypothesis implies that z is not a limit point of E, and so there is a radius $r = r_z$ say, such that $S(z, r_z) \cap E$ can contain no point of E except possibly z itself.

If we construct such a sphere about each point of $\bar{S}(0, R)$, we obtain an open cover of $\bar{S}(0, R)$, and compactness allows us to choose a finite subset which covers $\bar{S}(0, R)$, and so certainly covers E. But each element of this finite subcover can contain at most one point of E, and it follows that E must be finite.

A consequence of the Bolzano–Weierstrass property is that if E is an infinite subset of a compact set K, then E must have a limit point in K. For K, being compact, must be bounded, so E is bounded and must have a limit point which will also be a limit point of K. But K is closed, and so the limit point is in K.

(B) (Separation property). *Let K be a compact set, F be closed, and $K \cap F = \varnothing$. Then there is a $\delta > 0$, for which $z_1 \in K$, $z_2 \in F$ implies $|z_1 - z_2| \geqslant \delta$.*

To establish this, let z be any point of K. Since F is closed, there is a radius $r = r_z > 0$ for which $S(z, r_z) \cap F = \varnothing$. The spheres

$$\{S(z, \tfrac{1}{2}r_z): z \in K\}$$

form an open cover of K, and so by compactness we can choose a finite subcover, say corresponding to points $z_1, z_2, \ldots, z_k$, so that

$$K \subset \bigcup_{i=1}^{k} (S(z_i, \tfrac{1}{2}r_{z_i})).$$

Let $\delta = \min(\frac{1}{2}r_{z_1}, \frac{1}{2}r_{z_2}, \ldots, \frac{1}{2}r_{z_k})$.

Then for any point $z \in K$, $z \in S(z_j, \frac{1}{2}r_{z_j})$ for some choice of j, $1 \leqslant j \leqslant k$, while if $z' \in F$, $|z_j - z'| > r_{z_j}$. Hence

$$|z - z'| \geqslant |z' - z_j| - |z - z_j| > r_{z_j} - \tfrac{1}{2}r_{z_j} = \tfrac{1}{2}r_{z_j} \geqslant \delta$$

as required.

(vi) For any set E, the closure of E (which we will denote by $\bar{E}$) is the set made up of E together with its limit points. If z is not in $\bar{E}$, then (ii) shows that there is a sphere $S(z, r)$ which contains no points of E. The fact that $S(z, r)$ is an open set shows that it cannot contain any limit points of E, and so $S(z, r) \cap \bar{E} = \varnothing$. We have shown that the complement of $\bar{E}$ is open, and so that $\bar{E}$ is always a closed set. In particular it is easily seen that for any open sphere $S(a, r)$, the points of $S(a, r)$ and $C(a, r)$ are the limit points of $S(a, r)$, and so the closed sphere $\bar{S}(a, r)$ is the closure of the corresponding open sphere.

The boundary (denoted ∂E) of a set E is the set of all points which are limit points of both E and its complement.

The interior (denoted E°) of a set E comprises these points which are the centre of an open sphere contained in E: equivalently $E^\circ = \bar{E} \setminus \partial E$.

$\bar{E}$ is the smallest closed set containing E, while E° is the largest open set contained in it. We say that a set $E \subset F$ is dense in F if $\bar{E} \supset F$.

Having described the properties of subsets of **C**, we now investigate functions or mappings of one subset to another. We assume a familiarity with the set theoretic terminology concerning functions (range, domain, one to one, onto, etc.).

Definition 1.3. Let $E \subset G$ and f be a mapping of E into **C**.

We say f is continuous on E if for each $a \in E$ and each $\varepsilon > 0$, there is a $\delta = \delta(a, \varepsilon)$, such that $|z - a| < \delta$ and $z \in E$ imply

$$|f(z) - f(a)| < \varepsilon.$$

Equivalently, f maps $S(a, \delta) \cap E$ into $S(f(a), \varepsilon)$.

We say f is uniformly continuous on E if for each $\varepsilon > 0$, there is a $\delta = \delta(\varepsilon)$ depending on ε only, such that $z, z' \in E$, and $|z - z'| < \delta$ imply $|f(z) - f(z')| < \varepsilon$.

Theorem 1.4. *Let K be compact, and f a continuous mapping of K into* **C**. *Then*

(*i*) *f is uniformly continuous,*
(*ii*) *the range f(K) of f is compact, and*
(*iii*) *if in addition f is one-to-one, then f^{-1} (the inverse mapping of f) is also continuous, and so uniformly continuous by* (i) *and* (ii).

Proof. (i) Let $\varepsilon > 0$ be given, and for each $z \in K$, choose $r = r_z > 0$, such that f maps $S(z, r_z)$ into $S(f(z), \frac{1}{2}\varepsilon)$. The spheres

$$\{S(z, \tfrac{1}{2}r_z): z \in K\}$$

form an open cover of K: we use compactness to choose a finite subcover, corresponding to points $z_1, z_2, \ldots, z_k$ say. We then choose $\delta = \min(\frac{1}{2}r_{z_1}, \frac{1}{2}r_{z_2}, \ldots, \frac{1}{2}r_{z_k})$, and show that it has the required property.

Suppose that $z, z' \in E$ and $|z - z'| < \delta$. Then for some z_j $(j = 1, 2, \ldots, k)$, $z \in S(z_j, \frac{1}{2}r_{z_j})$, and since $|z - z'| < \delta \leqslant \frac{1}{2}r_{z_j}$, we have that both z and z' belong to $S(z_j, r_{z_j})$. Hence both $f(z)$ and $f(z')$ belong to $S(f(z_j), \frac{1}{2}\varepsilon)$, and so

$$|f(z) - f(z')| \leqslant |f(z) - f(z_j)| + |f(z_j) - f(z')| < \tfrac{1}{2}\varepsilon + \tfrac{1}{2}\varepsilon = \varepsilon.$$

(ii) To show $f(K)$ is compact, we consider an arbitrary open cover (G_α) (where α runs through some unspecified indexing set) for $f(K)$. For each α, denote by H_α the set $f^{-1}(G_\alpha)$ which is the set of points which are mapped by f into G_α. Since the (G_α) cover $f(K)$, the sets (H_α) form a cover of K. Suppose then that $z \in K$, and so $z \in H_{\alpha_1}$ for a suitable choice of α_1. Then $f(z) \in G_{\alpha_1}$ which is open, and so for some $r > 0$, $S(f(z), r) \subset G_{\alpha_1}$. The continuity of f shows that there is a $\delta > 0$ such that f maps $S(z, \delta) \cap K$ into $S(f(z), r)$, and so into G_{α_1}. Hence $S(z, \delta) \cap K \subset H_{\alpha_1}$, and so if H'_{α_1} denotes the union of all the $S(z, \delta)$ constructed from points $z \in H_{\alpha_1}$, then H'_{α_1} is open in **C**, and $H'_{\alpha_1} \cap K = H_{\alpha_1}$.

We have shown that the sets (H'_α) form an open cover of K and so since K is compact, that we can choose a finite subcover of K. Then the sets $(G_\alpha) = (f(H'_\alpha \cap K))$ which correspond to this finite subcover form a finite subcover of $f(K)$, and it follows that $f(K)$ is compact.

(iii) We suppose that f is one to one, so that a well-defined inverse function f^{-1} exists. Let b be any point of $f(K)$, and let $a = f^{-1}(b)$.

Suppose $\varepsilon > 0$ is given: to show f^{-1} is continuous we have to show that there is a $\delta > 0$ such that f^{-1} maps $S(b, \delta) \cap f(K)$ into $S(a, \varepsilon)$.

Now $S(a, \varepsilon)$ is open, and so $K \setminus S(a, \varepsilon)$ is a closed subset of K, which is compact. This now implies that $K' = K \setminus S(a, \varepsilon)$ is also compact (for any open cover of K' extends to an open cover of K on the addition of $S(a, \varepsilon)$: a finite subcover chosen to cover K will then also cover K' and $S(a, \varepsilon)$ may then be dropped without affecting this property). Hence by (ii) $f(K')$ is compact, and therefore closed, and the one to one property of f shows that $b \notin f(K')$. Hence there is a $\delta > 0$, with $S(b, \delta) \cap f(K') = \varnothing$, and so $f^{-1}(S(b, \delta)) \cap K' = \varnothing$.

Hence $f^{-1}(S(b, \delta)) \subset S(a, \varepsilon)$ as required.

From (ii) of this theorem it follows that if f is real valued (i.e. $f: K \to \mathbf{R}$), then the range of f is bounded and closed—in particular the infimum and supremum of the range of f are attained at points of K.

Definition 1.5. Suppose $E \subset \mathbf{C}$, and that f, $(f_n)_{n=1}^{\infty}$ are functions from E to $\mathbf{C}$.

We say (f_n) converges to f pointwise on E as $n \to \infty$, if for each $z \in E$, and $\varepsilon > 0$, there is an $n_0 = n_0(z, \varepsilon)$ such that

$$|f_n(z) - f(z)| < \varepsilon$$

if $n \geqslant n_0$.

We say (f_n) converges to f uniformly on E as $n \to \infty$, if for each $\varepsilon > 0$, there is an $n_0 = n_0(\varepsilon)$ such that $|f_n(z) - f(z)| < \varepsilon$ for all $z \in E$, if $n \geqslant n_0$. We say that a series converges uniformly to a sum s if the partial sums converges uniformly to s according to this definition.

Theorem 1.6. (i) *Suppose that the sequence (f_n) converges to f uniformly on E, and that each f_n is continuous on E. Then f is continuous on E.*

(ii) *Let $(a_n)_{n=0}^{\infty}$ be a sequence of complex numbers, and define R by the formula $R^{-1} = \limsup |a_n|^{1/n}$, if this limit superior is a real number in $(0, \infty)$. If the limit is infinite we take $R = 0$, while if the limit is zero we write $R = +\infty$. R is called the radius of convergence of the power series which we now consider.*

Let a be any complex number.

Then the power series $\sum_{n=0}^{\infty} a_n(z-a)^n$ has the following properties;

(a) *if $R > 0$, the series is absolutely convergent at every point of*

$S(a, R)$ and uniformly convergent on $\bar{S}(a, r)$ if $0 < r < R$ (in particular if $R = +\infty$, the series is absolutely convergent at every point of **C**, *and uniformly convergent on every bounded set).*

(b) *if $R < \infty$, the series diverges at every point not in $\bar{S}(a, R)$.*

Proofs. (i) is analogous to (and slightly simpler than) the first part of theorem A.13 in Appendix A to which the reader is referred for details.

(ii) is an immediate consequence of the Cauchy nth root test for convergence of a series with positive terms, together with the Weierstrass M-test for uniform convergence. Notice that no assertion is made as to the behaviour of the series on $C(a, R)$.

Definition 1.7. The exponential function, denoted by $\exp(z)$ or e^z, is defined by the equation

$$\exp z = \sum_{n=0}^{\infty} \frac{z^n}{n!}.$$

By (ii) of 1.6, the series is absolutely convergent for all z, and uniformly convergent on $S(0, R)$ for any $R > 0$. Combining this with (i) of 1.6 we see that exp is a continuous function from **C** to itself. Further properties of this most important of all special functions are given in 1.8 below, and in 1.16 (iv).

Theorem 1.8. *The exponential function has the following properties;*

(a) $e^{z+w} = e^z \cdot e^w$ *for all* $z, w \in \mathbf{C}$, $e^0 = 1$, *and* $e^z \neq 0$ *for any* $z \in \mathbf{C}$.

(b) *For real x, the function $x \to e^x$ which maps* **R** *to itself is monotone increasing, and strictly positive. For large x, e^x tends to infinity faster than any positive power of x;*

$$\text{if} \quad k > 0, \quad \text{then} \quad x^{-k}e^x \to +\infty \quad \text{as} \quad x \to +\infty.$$

(c) *There is a number* $\pi = 3{\cdot}14159\ldots$ *such that* $\exp[i(\pi/2)] = i$, *and* $e^z = 1$ *if and only if* $z = 2k\pi i$ *for some integer k.*

(d) *For real x, the function $x \to e^{ix}$ maps* **R** *onto $C(0, 1)$, and is one-to-one on each interval of the form $[\theta, 2\pi + \theta)$, $\theta \in \mathbf{R}$.*

(e) *The trigonometric functions* sin *and* cos *are defined for complex z by*

$$\cos z = \tfrac{1}{2}(e^{iz} + e^{-iz}) = \sum_{k=0}^{\infty} (-1)^k z^{2k}/(2k)!$$

$$\sin z = \frac{1}{2i}(e^{iz} - e^{-iz}) = \sum_{k=0}^{\infty} (-1)^k z^{2k+1}/(2k+1)!$$

where the power series expansions are absolutely convergent at all points of **C**. *In particular, for real* x,

$$\cos x = \operatorname{Re}(e^{ix}), \qquad \sin x = \operatorname{Im}(e^{ix}).$$

(f) *The hyperbolic functions* sinh *and* cosh *are defined by*

$$\cosh z = \tfrac{1}{2}(e^{z} + e^{-z}), \qquad \sinh z = \tfrac{1}{2}(e^{z} - e^{-z}),$$

and so for all $z \in \mathbf{C}$, *we have*

$$\cosh(iz) = \cos z, \qquad \sinh(iz) = i \sin z.$$

Proofs of these properties may be found in the book by W. Rudin, mentioned in the bibliography.

§2. DIFFERENTIATION AND INTEGRATION OF COMPLEX FUNCTIONS

Definition 1.9. Let f be a complex valued function defined on a set $E \subset \mathbf{C}$. Let c be a point of E which is also a limit point of E.

(i) We say f is real differentiable at c relative to E if there are complex constants A and B for which if $z \in E$,

$$f(z) = f(c) + A(x - a) + B(y - b) + \varepsilon|z - c|,$$

(where we have written $z = x + iy$, $c = a + ib$), and $\varepsilon \to 0$ as $|z - c| \to 0$.

(ii) We say f is complex differentiable at c relative to E if there is a complex constant C, for which if $z \in E$,

$$f(z) = f(c) + C(z - c) + \varepsilon'|z - c|,$$

and $\varepsilon' \to 0$ as $|z - c| \to 0$.

Remarks 1.10. Part (i) of the above definition is equivalent to the usual definition of differentiability for a function of two real variables. The constants A and B are simply the partial derivatives $\partial F/\partial x$, $\partial F/\partial y$ (evaluated at c) where $f(x + iy) = F(x, y)$. In particular this definition involves four real constants (the real and imaginary parts of A and B).

Notice also that if E is an interval $[a, b]$ in $\mathbf{R}$, then both (i) and (ii) reduce to the usual definition of a differentiable function of one real variable.

Part (ii) of the definition is satisfied if and only if

$$|\varepsilon'| = \left|\frac{f(z) - f(c)}{z - c} - C\right| \to 0 \quad \text{as} \quad z \to c,$$

and so is the most natural extension to complex numbers of the idea of a derivative as the limit of a 'difference quotient'.

Notice that (ii) involves only two real constants (the real and imaginary parts of C), and so this is a more restrictive condition than (i). We call the constant C the (complex) derivative of f at c, and denote it by $f'(c)$. We shall normally apply these definitions only when the set E on which f is defined is an open set in C, in which case all points z with $|z - c|$ sufficiently small are in E.

In this case the two definitions are related in the following way.

Theorem 1.11 (Cauchy–Riemann equations). *Suppose f is defined in an open set E, and that for $z = x + iy \in E$, we write*

$$f(x + iy) = u(x, y) + iv(x, y),$$

where u and v are real valued functions of the real variables x and y.

Then f is complex differentiable at $c = a + ib \in E$ if and only if

(i) *u and v are both real differentiable at (a, b), and*

(ii) $(\partial u/\partial x)(a, b) = (\partial v/\partial y)(a, b)$, *and*

$$(\partial u/\partial y)(a, b) = -(\partial v/\partial x)(a, b)$$

Proof. We suppose throughout that $|z - c|$ is taken small enough to ensure $z \in E$.

Suppose then that f is complex differentiable at c.

Then we may write

$$f(z) = f(c) + C(z - c) + \varepsilon'|z - c|,$$

where $\varepsilon' \to 0$ as $z \to c$.

If we write $C = C_1 + iC_2$ and $\varepsilon' = \varepsilon_1' + i\varepsilon_2'$, where $C_1, C_2, \varepsilon_1', \varepsilon_2'$, are real, then on equating real and imaginary parts we obtain

$$u(x, y) = u(a, b) + C_1(x - a) - C_2(y - b) + \varepsilon_1'|z - c|,$$

and (*)

$$v(x, y) = v(a, b) + C_2(x - a) + C_1(y - b) + \varepsilon_2'|z - c|,$$

where $\varepsilon_1', \varepsilon_2' \to 0$ as $|z - c| \to 0$. Hence u, v are real differentiable,

$$\frac{\partial u}{\partial x}(a, b) = C_1 = \text{Re} f'(c) = \frac{\partial v}{\partial y}(a, b),$$

and

$$\frac{\partial u}{\partial y}(a, b) = -C_2 = -\text{Im} f'(c) = -\frac{\partial v}{\partial x}(a, b).$$

Conversely suppose u and v satisfy conditions (i) and (ii) of the theorem.

Then if we write C_1 for the common value of $\partial u/\partial x$ and $\partial v/\partial y$, and $-C_2$ for $-\partial u/\partial y = \partial v/\partial x$, the equations (*) will be satisfied for some (real) ε_1, ε_2 which tend to zero with $|z - c|$.

On combining these together we obtain

$$\begin{aligned} f(z) &= u(x, y) + v(x, y) \\ &= u(a, b) + iv(a, b) + (C_1 + iC_2)((x - a) + i(y - b)) \\ &\quad + (\varepsilon_1 + i\varepsilon_2)|z - c| \\ &= f(c) + (C_1 + iC_2)(z - c) + \varepsilon|z - c|, \end{aligned}$$

where $\varepsilon = \varepsilon_1 + i\varepsilon_2 \to 0$ with $|z - c|$.

Hence f is complex differentiable at c, with $f'(c) = C_1 + iC_2$.

From now until Chapter 6 we shall work entirely with complex differentiable functions, and in order to abbreviate the rather cumbersome terminology, we make the following definition.

Definition 1.12. Let f be defined on an open set $G \subset \mathbf{C}$. We say f is regular on G if f is complex differentiable at each point of G. We say f is regular at a point a if in fact f is regular in an open set to which a belongs. Any point at which f is not regular is called a singularity of f. (Notice that this very sweeping definition of a singularity includes in particular boundary points of G where f may not even be defined!)

Exercise 5 at the end of this chapter illustrates these ideas in more detail.

Part (ii) of the following theorem gives (in a sense which will be made more precise in Chapter 3) the most general example of a regular function.

Theorem 1.13.
(i) *Any polynomial* $P(z) = a_0 + a_1 z + a_2 z^2 + \cdots + a_n z^n$ *is regular on* $\mathbf{C}$, *with* $P'(z) = a_1 + 2a_2 z + \cdots + n a_n z^{n-1}$.
(ii) *Suppose the power series* $\sum_{n=0}^{\infty} a_n (z-a)^n$ *has radius of convergence* $R > 0$. *Then the function* f *defined in* $S(a, R)$ *by*

$$f(z) = \sum_{n=0}^{\infty} a_n (z-a)^n$$

is regular in $S(a, R)$, *and its derivative is given by*

$$f'(z) = \sum_{n=1}^{\infty} n a_n (z-a)^{n-1}.$$

(*In the case* $R = \infty$, *take* $\mathbf{C}$ *for* $S(a, R)$.)

Proof. Evidently it is sufficient to prove (ii), since (i) is a special case of it. Without loss of generality, we take $a = 0$.

The power series $\sum_{n=1}^{\infty} n a_n z^{n-1}$ has radius of convergence given by $R' = (\limsup |n a_n|^{1/(n-1)})^{-1}$, and since $n^{1/n} \to 1$ as $n \to \infty$, it follows that $R' = R$.

Hence we may define a function g on $S(0, R)$ by

$$g(z) = \sum_{n=1}^{\infty} n a_n z^{n-1}.$$

Given $z \in S(0, R)$ choose r with $|z| < r < R$, and suppose $|w| < r$ also. Then we may calculate the derivative of f as follows.

$$\begin{aligned}
\frac{f(w) - f(z)}{w - z} &- g(z) \\
&= \sum_{n=1}^{\infty} \left(a_n \frac{w^n - z^n}{w - z} - n a_n z^{n-1} \right) \\
&= \sum_{n=2}^{\infty} a_n \left(\sum_{m=0}^{n-1} z^{n-m-1} w^m - n z^{n-1} \right) \\
&= \sum_{n=2}^{\infty} a_n \sum_{m=1}^{n-1} (w^m - z^m) z^{n-m-1} \\
&= (w - z) \sum_{n=2}^{\infty} a_n \sum_{m=1}^{n-1} z^{n-m-1} (w^{m-1} + w^{m-2} z + \cdots + z^{m-1}).
\end{aligned}$$

$$\text{Hence } \left|\frac{f(w)-f(z)}{w-z} - g(z)\right| \leqslant |w-z| \sum_{n=2}^{\infty} |a_n| \left(\sum_{m=1}^{n-1} m\right) r^{n-2}$$
$$= \tfrac{1}{2}|w-z| \sum_{n=2}^{\infty} n(n-1)|a_n| r^{n-2},$$

a series which is again convergent by the nth root test since $r < R$.

Hence as $w \to z$,

$$\frac{f(w)-f(z)}{w-z} - g(z) \to 0,$$

and so $f'(z)$ exists and is equal to $g(z)$ at all points of $S(0, R)$.

Corollary 1.14. *Suppose as in* (ii) *of* 1.13 *that* $f(z) = \sum_{n=0}^{\infty} a_n(z-a)^n$, *where the series is convergent on* $S(a, R)$ *with* $R > 0$.

Then for each $k = 1, 2, 3, \ldots, f$ *is* k *times differentiable, and if* $z \in S(a, R)$,

$$f^{(k)}(z) = \sum_{n=k}^{\infty} n(n-1)\ldots(n-k+1)a_n(z-a)^{n-k}.$$

In particular $f^{(k)}(a) = k!a_k$, *so that* f *may be written*

$$f(z) = \sum_{n=0}^{\infty} \frac{f^{(n)}(a)}{n!}(z-a)^n.$$

Proof. We use induction on k, and 1.13 gives us the case $k = 1$.

Suppose then that for some k we have shown that

$$f^{(k)}(z) = \sum_{n=k}^{\infty} n(n-1)\ldots(n-k+1)a_n(z-a)^{n-k},$$

where the series converges on $S(a, R)$. We apply 1.13 to this series to obtain

$$f^{(k+1)}(z) = \sum_{n=k+1}^{\infty} n(n-1)\ldots(n-k+1)(n-k)a_n(z-a)^{n-k-1},$$

which is the same formula with $k + 1$ in place of k. This establishes the result.

It should be emphasized that we have proved the result that f has derivatives of all orders and is the sum of the series

$$\sum_{n=0}^{\infty} \frac{f^{(n)}(a)}{n!}(z-a)^n$$

only for functions which are given initially by power series. These facts remain true, at any rate in suitable discs $S(a, R)$, for all regular functions, as theorem 3.1 will show.

We now turn to the subject of integration. Like complex differentiation, integration in the complex plane has a rather different 'flavour' from that in the real case. For integrals this is largely due to the fact that one has to integrate along paths which may be very complicated, in place of the usual line segments $[a, b]$ in the real case.

We begin then with a study of curves in the complex plane.

Definition 1.15. (i) A curve γ is a continuous mapping of a closed interval $[a, b] \subset \mathbf{R}$ into $\mathbf{C}$. $[a, b]$ is called the parameter interval of γ. Curves $\gamma_1: [a_1, b_1] \to \mathbf{C}$ and $\gamma_2: [a_2, b_2] \to \mathbf{C}$ are said to be equivalent if there is a continuous strictly increasing function

$$\phi: [a_1, b_1] \to [a_2, b_2]$$

for which $\gamma_1 = \gamma_2 \circ \phi$. It is an easy exercise to show that this notion of (curve) equivalence is an equivalence relation in the set theoretic sense. In particular, taking $\phi(t) = (t - a_1)/(a_2 - a_1)$, every curve is equivalent to one whose parameter interval is $[0, 1]$, and we shall generally (though not exclusively) assume that our curves are defined on $[0, 1]$.

(ii) For any curve γ, the range of γ is a subset of $\mathbf{C}$, called the track of γ, and denoted by γ^*,

$$\gamma^* = \{z : z = \gamma(t) \quad \text{for some} \quad t \in [0, 1]\}.$$

γ^* is compact by 1.4 (ii).

The points $\gamma(0)$ and $\gamma(1)$ of γ^* are called the initial and final points of γ respectively, and we say that γ is closed if they coincide.

(iii) Suppose that γ, δ are curves with $\gamma(1) = \delta(0)$. Then we can unite them to form a new curve (which we will denote by $\gamma \cup \delta$) by defining

$$\begin{aligned}(\gamma \cup \delta)(t) &= \gamma(2t) && \text{if} \quad 0 \leqslant t \leqslant \tfrac{1}{2} \\ &= \delta(2t - 1) && \text{if} \quad \tfrac{1}{2} \leqslant t \leqslant 1.\end{aligned}$$

The track $(\gamma \cup \delta)^*$ of $\gamma \cup \delta$ is thus simply $\gamma^* \cup \delta^*$.

For any curve γ we may define its 'opposite' $(-\gamma)$ by

$$(-\gamma)(t) = \gamma(1 - t), \qquad 0 \leqslant t \leqslant 1.$$

$(-\gamma)$ has initial point $\gamma(1)$ and final point $\gamma(0)$, and $(-\gamma)^* = \gamma^*$.

The definitions (ii) and (iii) are the same for any curve γ' equivalent to γ.

Definition 1.16. (i) Let γ be a curve, γ: $[a, b] \to \mathbf{C}$. If the derivative γ' exists and is continuous throughout $[a, b]$, then we say γ is smooth on $[a, b]$. In the definition of equivalence for smooth curves, we require that the function ϕ: $[a_1, b_1] \to [a_2, b_2]$ shall have a continuous derivative which is everywhere strictly positive. Normally the curves over which we shall integrate will be smooth except for a finite number of points at which γ may have a 'corner'. This is allowed for in the next definition.

(ii) A curve γ is called a path if there are a finite number of smooth curves $\gamma_1, \gamma_2, \ldots, \gamma_n$, with $\gamma = \gamma_1 \cup \gamma_2 \cup \cdots \cup \gamma_n$. Another way of describing this is to say that γ is a continuous function on $[a, b] \to \mathbf{C}$, and that there are points $t_0, t_1, \ldots, t_n$, with

$$a = t_0 < t_1 < \cdots < t_n = b,$$

such that γ has a derivative at each point $t \neq t_j$ $(j = 0, 1, \ldots, n)$, a left derivative at t_j $(j = 1, 2, \ldots, n)$ and a right derivative at t_j

$$(j = 0, 1, \ldots, n - 1).$$

Further we require that γ' shall be continuous when restricted to the closed intervals $[t_{j-1}, t_j]$$(j = 1, 2, \ldots, n)$, but that the left- and right-hand derivatives at $t_1, t_2, \ldots, t_{n-1}$ need not coincide. In this case we shall also say that γ is piecewise continuously differentiable.

Since γ' is continuous on the intervals $[t_{j-1}, t_j]$, we may integrate $|\gamma'|$ as a function of one real variable (for instance by the technique outlined in Appendix A) to obtain a positive real number which we call the length of γ.

$$L(\gamma) = \int_0^1 |\gamma'(t)|dt.$$

If γ_1 and γ_2 are equivalent, $L(\gamma_1) = L(\gamma_2)$.

(iii) For points $c, d \in \mathbf{C}$, the directed line segment $[c, d]$ from c to d is the path given by $\gamma(t) = c + t(d - c)$, $0 \leqslant t \leqslant 1$.

It is customary to use the notation $[c, d]$ also for the track of this curve. However, it should be remembered that $[c, d]$ is directed from c to d, and so $[d, c] = -[c, d]$ in the sense of 1.15(iii).

If $\gamma(t) = c + t(d - c)$, $\gamma'(t) = d - c$, and so

$$L(\gamma) = \int_0^1 |\gamma'(t)|dt = |d - c| \int_0^1 dt = |d - c|,$$

a result which should surprise no one!

A path of the form $[z_0, z_1] \cup [z_1, z_2] \cup \cdots \cup [z_{n-1}, z_n]$ is called a polygonal path from z_0 to z_n. Its length is $\sum_{j=1}^{n} |z_j - z_{j-1}|$. The set $E \in \mathbf{C}$ is said to be starred with respect to $a \in E$, if for all $z \in E$, the line segment $[a, z]$ is contained in E.

(iv) It follows from the properties of the exponential function, listed under theorem 1.8, parts (c) and (d), that if $a \in \mathbf{C}$, and $r > 0$, the curve given by $\gamma(t) = a + r\, e^{2\pi i t}$, $0 \leqslant t \leqslant 1$ is a closed path whose track is the circumference $C(a, r)$, and whose length is $2\pi r$.

Here again it is customary to identify the curve with its track; however if one uses $C(a, r)$ to denote the above curve, it is essential to use the function $\gamma(t) = a + r\, e^{2\pi i t}$ (or one equivalent to it). The curve $(-\gamma)(t) = a + r\, e^{-2\pi i t}$, $0 \leqslant t \leqslant 1$, has the same track but is not equivalent.

Our convention regarding circular paths then, is that they must always be parametrized by the function $\gamma(t) = a + r\, e^{2\pi i t}, 0 \leqslant t \leqslant 1$, and since

$$\gamma(0) = \gamma(1) = a + r, \quad \gamma(\tfrac{1}{4}) = a + ir,$$
$$\gamma(\tfrac{1}{2}) = a - r \quad \text{and} \quad \gamma(\tfrac{3}{4}) = a - ir$$

we see that this convention is equivalent to always choosing the anti-clockwise direction on the circle. This observation is of great importance in lemma 1.22, and again in theorem 5.7.

With these notions concerning paths in $\mathbf{C}$, we can define what is meant by saying that a set is connected, and give several equivalent descriptions.

Definition 1.17. Let G be an open set in $\mathbf{C}$. We say that G is connected if it has any one (and so, as a consequence of the proof of equivalence which follows, all three) of the following properties:

(a) Every decomposition of G into a union of two disjoint open subsets G_1 and G_2 must be trivial; that is G_1 or G_2 must be empty.

(b) Every pair of points in G are the initial and final points of a polygonal path whose track lies in G.

(c) Every continuous function from G to $\mathbf{Z}$ (the integers) is constant.

The definition of connectedness for general (non-open) sets is discussed in exercise 1.16.

Proof of equivalence. *(a)* $\Rightarrow$ *(b)* Let z_0 be any point of G: we have to use (a) to prove the existence of a polygonal path in G from z_0 to any other point of G.

Let G_1 be the set of points of G which can be reached from z_0 by a polygonal path lying in G, and let $G_2 = G \setminus G_1$. We prove that both G_1 and G_2 are open, and hence since $G_1 \neq \varnothing$ (it contains z_0!), (a) shows that $G_2 = \varnothing$ as required.

To show G_1 is open, let w_1 be any point of G_1. The definition of G_1 shows that there is a polygonal path γ from z_0 to w_1. Since $w_1 \in G_1 \subset G$ which is open in $\mathbf{C}$, there is an $r > 0$, with $S(w_1, r) \subset G$. Then if w is any point of $S(w_1, r)$, the segment $[w_1, w]$ lies in G, and so $\gamma \cup [w_1, w]$ is a polygonal path in G from z_0 to w. Hence $S(w_1, r) \subset G_1$, so that G_1 is open.

Now let w_2 be any point of G_2. Again, since G is open, there is a radius $r' > 0$, such that $S(w_2, r') \subset G$. Suppose that a point w' is in $G_1 \cap S(w_2, r')$. Then there is a polygonal path γ' in G, from z_0 to w', and so $\gamma' \cup [w', w_2]$ is a polygonal path in G from z_0 to w_2, contrary to the fact that $w_2 \in G_2$. Hence no point of $S(w_2, r')$ can be in G_1, so $S(w_2, r') \subset G_2$, proving that G_2 is open, as required.

(b) $\Rightarrow$ *(c)* Let g be a continuous function on G to $\mathbf{Z}$, and let a, b be arbitrary points of G.

There exists by (b) a polygonal path γ: $[0, 1] \rightarrow G$, with $\gamma(0) = a$ and $\gamma(1) = b$. Then $f \circ \gamma$ is a continuous function from $[0, 1]$ to $\mathbf{Z}$. The intermediate value property for continuous functions on a real interval now shows that $f \circ \gamma$ must be constant, and in particular that $f(a) = f(\gamma(0)) = f(\gamma(1)) = f(b)$. Hence f is constant on G.

(c) $\Rightarrow$ *(a)* We prove this implication by a 'contrapositive' argument: that is we assume (a) fails, and show that in this case, (c) fails also.

If (a) does not hold, we can write $G = G_1 \cup G_2$, where G_1 and G_2 are disjoint, non-empty open subsets of $\mathbf{C}$. We define f on G by

$$f(z) = \begin{cases} 1 & \text{if } z \in G_1, \\ 2 & \text{if } z \in G_2 \end{cases}$$

which gives a non-constant function $G \to \mathbf{Z}$. f is easily seen to be continuous on G, since for instance if $z \in G_1$ and $\varepsilon > 0$ is given, then for some $\delta > 0$, $S(z, \delta) \subset G_1$, and so

$$|f(z') - f(z)| = 0 < \varepsilon \quad \text{if} \quad |z - z'| < \delta.$$

We now define, for an arbitrary open set G, a decomposition into connected subsets called the components of G.

Definition 1.18. Let G be an open subset of $\mathbf{C}$. The binary relation '$\sim$' defined for elements $z, z' \in G$ by

$$z \sim z' \quad \text{if there is a polygonal path in } G \text{ from } z \text{ to } z'$$

is evidently an equivalence relation (that is, it is reflexive, symmetric, and transitive) on G.

The equivalence classes into which G is decomposed by this relation are disjoint open connected subsets of G which are called the connected components of G. The existence of a denumerable dense subset of $\mathbf{C}$, (for instance the points whose real and imaginary parts are rational numbers) each point of which can lie in at most one component, and such that each component contains at least one point (in fact denumerably many!), shows that the number of components of an open set is either finite or denumerably infinite.

For examples of the decomposition of open sets into their components, the reader can consult exercise 12 at the end of this chapter and also examples 5.2.

Definition 1.19. Let γ be a path in $\mathbf{C}$, and f a continuous function from γ^* to $\mathbf{C}$. Then we define the integral of f over γ by

$$\int_\gamma f(z)\, dz = \int_0^1 f(\gamma(t))\gamma'(t)\, dt.$$

Notice that since γ is a path, γ' is piecewise continuous, and so the right-hand side may be evaluated by any process which will integrate such functions (one such process is discussed in Appendix A).

It is important to realize that the value of the integral depends on the path (that is, the mapping $[0, 1] \to \mathbf{C}$) rather than simply the track γ^*. In the very important special case when γ is $C(a, r)$, the convention of 1.16 (iv) applies, and we have

$$\int_{C(a,r)} f(z)\,dz = \int_0^1 f(a + r\,e^{2\pi it})r2\pi i\,e^{2\pi it}\,dt$$
$$= 2\pi ir \int_0^1 f(a + r\,e^{2\pi it})\,e^{2\pi it}\,dt.$$

The principal formal properties of the integral are listed in theorem 1.20, while 1.22 comprises a simple calculation which is of crucial importance in the further development of the subject.

Theorem 1.20

(i) *If γ_1 and γ_2 are equivalent paths in* $\mathbf{C}$, *and f is continuous on $\gamma_1^*(=\gamma_2^*)$, then*

$$\int_{\gamma_1} f(z)\,dz = \int_{\gamma_2} f(z)\,dz.$$

(ii)
$$\int_{(-\gamma)} f(z)\,dz = -\int_{\gamma} f(z)\,dz.$$

(iii) *If the final point of γ_1 is the initial point of γ_2, so that $\gamma = \gamma_1 \cup \gamma_2$ is defined, then*

$$\int_{\gamma} f(z)\,dz = \int_{\gamma_1} f(z)\,dz + \int_{\gamma_2} f(z)\,dz.$$

(iv) *If f and g are continuous on γ^*, and α, β are complex constants, then*

$$\int_{\gamma} (\alpha f(z) + \beta g(z))\,dz = \alpha \int_{\gamma} f(z)\,dz + \beta \int_{\gamma} g(z)\,dz.$$

(v) *If f is continuous on γ^*, and F is complex differentiable at each point of γ^*, with $F'(z) = f(z)$ for each $z \in \gamma^*$, then*

$$\int_{\gamma} f(z)\,dz = F(\gamma(1)) - F(\gamma(0)).$$

The condition holds in particular if F is regular on an open set which contains γ^*, and $F' = f$ on γ^*.

(vi) *Suppose $|f(z)| \leqslant M$ for all $z \in \gamma^*$. Then*

$$\left|\int_{\gamma} f(z)\,dz\right| \leqslant ML(\gamma).$$

Proof. Parts (i)–(iv) are immediate from definition 1.19 and the corresponding properties of the integral over intervals in **R**. In particular (i) is a consequence of the rule for 'integration by substitution'.

To prove (v), suppose

$$0 = t_0 < t_1 < \cdots < t_{j-1} < t_j < \cdots < t_n = 1$$

and that γ is smooth on $[t_{j-1}, t_j] \subset [0, 1]$, $\quad j = 1, 2, \ldots, n$.

Then

$$\int_\gamma f(z)\,dz = \int_0^1 f(\gamma(t))\gamma'(t)\,dt = \sum_{j=1}^n \int_{t_{j-1}}^{t_j} f(\gamma(t))\gamma'(t)\,dt$$

$$= \sum_{j=1}^n \int_{t_{j-1}}^{t_j} F'(\gamma(t))\gamma'(t)\,dt = \sum_{j=1}^n \int_{t_{j-1}}^{t_j} (F \circ \gamma)'(t)\,dt$$

$$= \sum_{j=1}^n \{F(\gamma(t_j)) - F(\gamma(t_{j-1}))\} = F(\gamma(1)) - F(\gamma(0)).$$

Finally, for (vi), we have

$$\left|\int_\gamma f(z)\,dz\right| = \left|\int_0^1 f(\gamma(t))\gamma'(t)\,dt\right| \leqslant \int_0^1 |f(\gamma(t))|\,|\gamma'(t)|\,dt$$

$$\leqslant M \int_0^1 |\gamma'(t)|\,dt = ML(\gamma).$$

Examples 1.21

(i) Let γ be the closed path made up of the segment $[0, 1]$ followed by the perimeter of the closed unit circle from 1 to i followed by the segment $[i, 0]$.

We evaluate $\int_\gamma x^2\,dz$, where $x = \operatorname{Re} z$.

Firstly $\int_{[0,1]} x^2\,dz = \int_0^1 t^2\,dt = \frac{1}{3}$ since $[0, 1]$ is parametrized by $z(=x) = t$, $0 \leqslant t \leqslant 1$.

The value of the integral along the part of the unit circumference is found by putting $\gamma_1(t) = \exp\{\frac{1}{2}\pi i t\}$, $\quad 0 \leqslant t \leqslant 1$.

$$\int_{\gamma_1} x^2\,dz = \int_0^1 \cos^2(\tfrac{1}{2}\pi t)\tfrac{1}{2}\pi i\, e^{1/2\pi i t}\,dt$$

$$= i\left(\int_0^{\pi/2} \cos^2\theta(\cos\theta + i\sin\theta)\,d\theta\right) = i(\tfrac{2}{3} + \tfrac{1}{3}i) = \tfrac{2}{3}i - \tfrac{1}{3},$$

after an elementary calculation.

Finally on $[i, 0]$, $x = \operatorname{Re} z = 0$, so $\int_{[i,0]} x^2\,dz = 0$.

Hence $\int_\gamma x^2\,dz = \frac{1}{3} + (\frac{2}{3}i - \frac{1}{3}) + 0 = \frac{2}{3}i$.

(ii) Let γ be as in (i). We evaluate $\int_\gamma z^2\,dz$.

Now f given by $f(z) = z^2$ is regular in $\mathbf{C}$, so that we can apply (v) of 1.20. Then $\int_\gamma z^2\,dz = [\frac{1}{3}z^3]_{z=\gamma(0)}^{z=\gamma(1)} = 0$ since $\gamma(0) = \gamma(1) = 0$.

Lemma 1.22. Let γ be the circular path $C(a, r)$ given (as always) by $\gamma(t) = a + re^{2\pi it}$, $0 \leqslant t \leqslant 1$.

Then

$$\frac{1}{2\pi i}\int_\gamma \frac{dz}{z - w} = \begin{cases} 1 & \text{if } |w - a| < r, \\ 0 & \text{if } |w - a| > r. \end{cases}$$

Proof. Since we assume $|w - a| \neq r$, $w \notin C(a, r)$ so $z - w \neq 0$ if $z \in C(a, r)$ and the integral is well defined.

Then we have

$$\frac{1}{2\pi i}\int_\gamma \frac{dz}{z - w} = \frac{1}{2\pi i}\int_0^1 \frac{2\pi i r e^{2\pi it}\,dt}{a + r\,e^{2\pi it} - w} = \int_0^1 \frac{dt}{1 - b\,e^{-2\pi it}},$$

where we have written b for $(w - a)/r$.

The case $|w - a| < r$, corresponds to $|b| < 1$. In this case we expand $(1 - be^{-2\pi it})^{-1}$ in increasing powers of b, and obtain

$$\sum_{k=0}^{\infty} (b\,e^{-2\pi it})^k\,dt = \sum_{k=0}^{\infty} b^k\,e^{-2\pi ikt},$$

a series which converges absolutely and uniformly for $t \in [0, 1]$, by the Weierstrass M-test.

Hence we may integrate term by term to obtain

$$\frac{1}{2\pi i}\int_\gamma \frac{dz}{z - w} = \sum_{k=0}^{\infty} b^k \int_0^1 e^{-2\pi ikt}\,dt.$$

Now

$$\int_0^1 e^{-2\pi ikt}\,dt = \begin{cases} 1 & \text{if } k = 0, \\ -\dfrac{1}{2\pi ik}\,[e^{-2\pi ikt}]_{t=0}^{t=1} = 0 & \text{if } k \neq 0. \end{cases}$$

Hence only the term with $k = 0$ remains and its value is $b^0 = 1$.

In the case $|w - a| > r$, we have $|b| > 1$, and we expand in powers of b^{-1} to obtain

$$(1 - b\, e^{-2\pi it})^{-1} = -\frac{1}{b} e^{2\pi it}\left(1 - \frac{1}{b} e^{2\pi it}\right)^{-1}$$

$$= -\sum_{k=0}^{\infty} b^{-(k+1)} e^{2\pi i(k+1)t}.$$

Again we may integrate term by term, and in this case each term

$$\int_0^1 e^{2\pi i(k+1)t}\, dt = \frac{1}{2\pi i(k+1)} [e^{2\pi i(k+1)t}]_0^1 = 0.$$

In this case therefore, the integral has value zero.

We finish the introductory chapter with a result which shows that we may 'differentiate under the integral sign' just as in the usual real-variable case.

Theorem 1.23. *Let G be open in* **C**, *and f be a continuous function on $G \times [0, 1]$ to* **C**, *having the property that for each $t \in [0, 1]$, $f(z, t)$ is a regular function of $z \in G$. Denote the derivative of this regular function (that is, the partial derivative of f with respect to z) by f_1, and suppose that f_1 is also continuous on $G \times [0, 1]$.*

Then the function

$$F(z) = \int_0^1 f(z, t)\, dt, \qquad z \in G,$$

is regular on G, and its derivative is given by

$$F'(z) = \int_0^1 f_1(z, t)\, dt.$$

Proof. Let a be a given point of G, and having chosen a, take $R > 0$ such that $\bar{S}(a, R) \subset G$.

By hypothesis, f_1 is continuous on $G \times [0, 1]$, and is therefore uniformly continuous on $\bar{S}(a, R) \times [0, 1]$ by 1.4(i). Hence given $\varepsilon > 0$, there is a $\delta > 0$ such that $|f(z, t) - f(z', t')| < \varepsilon$ whenever $z, z' \in \bar{S}(a, R)$, $t, t' \in [0, 1]$ and $|z - z'|$ and $|t - t'|$ are both $< \delta$. Suppose then that $z \in \bar{S}(a, R)$, and consider $F(z) - F(a)$. We have

$$F(z) - F(a) = \int_0^1 \{f(z, t) - f(a, t)\}\, dt$$
$$= \int_0^1 \left\{\int_{[a,z]} f_1(w, t)\, dw\right\} dt.$$

Hence

$$\frac{F(z) - F(a)}{z - a} - \int_0^1 f_1(a, t)\, dt$$
$$= \int_0^1 \left\{\frac{1}{(z - a)} \int_{[a,z]} f_1(w, t)\, dw - f_1(a, t)\right\} dt$$
$$= \int_0^1 \left\{\frac{1}{(z - a)} \int_{[a,z]} (f_1(w, t) - f_1(a, t)\, dw\right\} dt.$$

If we now take $|z - a| < \delta$, then $|w - a| < \delta$ since $w \in [a, z]$. It follows that $|f_1(w, t) - f_1(a, t)| < \varepsilon$, and so if $|z - a| < \delta$ we may estimate the right-hand side of the above equation to obtain

$$\left|\frac{F(z) - F(a)}{z - a} - \int_0^1 f_1(a, t)\, dt\right| \leqslant \int_0^1 \frac{1}{|z - a|}\, \varepsilon |z - a|\, dt = \varepsilon.$$

Hence $F'(a) = \int_0^1 f_1(a, t)\, dt$, and the result follows, since a was an arbitrary point of G.

EXERCISES FOR CHAPTER 1

1. (i) Let z_1, z_2 be complex numbers with $z_1 + z_2 = 1$.

Show that $1 \leqslant |z_1| + |z_2|$ with equality if and only if z_1 and z_2 are real and positive.

By considering $z_1/(z_1 + z_2)$ and $z_2/(z_1 + z_2)$ (and also the case $z_1 + z_2 = 0$!) deduce the general case of the triangle inequality, and also the condition for equality.

(ii) Prove that for any complex numbers z, w,

$$||z| - |w|| \leqslant |z - w|.$$

What are the conditions for equality?

2. Which of the following sets are (a) open, (b) closed, (c) compact and, if open (d) connected?

(i) $\{z\colon \operatorname{Re} z > 0\}$,
(ii) $S(-1, 1) \cup \bar{S}(1, 1)$,
(iii) $\{z\colon |\operatorname{Re} z| + |\operatorname{Im} z| \leqslant 1\}$,
(iv) $\{z\colon |z - 1| > 2\}$,
(v) $C(0, 1) \cup [1, 2]$,
(vi) $\{z\colon \operatorname{Re} z > 0 \text{ or } \operatorname{Im} z \neq 0\}$.

3. Describe the closure of each of the sets in exercise 2.

4. Using the properties of the exponential function listed in 1.8, describe the set of all solutions to the equations (i) $e^z = -1$, (ii) $\sin z = 1$, (iii) $\cos z = 0$, (iv) $\tan z (= \sin z/\cos z) = 1$.

5. Describe the sets of points at which the following functions are complex differentiable, and the (open!) sets on which they are regular.

(i) $f(z) = xy + \frac{1}{2}i(x^2 - y^2), \quad z = x + iy \in \mathbf{C}.$
(ii) $f(z) = \sin x \cosh y + i \cos x \sinh y, \quad z = x + iy \in \mathbf{C}.$
(iii) $f(z) = x^2 + iy^2, \quad z = x + iy \in \mathbf{C}.$
(iv) $f(z) = \exp(-z^{-2}), \quad z \neq 0, \quad f(0) = 0.$

6. Find the radii of convergence of the following series:

(i) $\sum_{n=1}^{\infty} \frac{z^n}{n^2}$, (ii) $\sum_{n=0}^{\infty} \frac{(n+1)!}{(2n)!} z^n$, (iii) $\sum_{n=0}^{\infty} 2^n z^{n^2}$.

(iv) $\sum_{n=1}^{\infty} (n^{1/n} - 1) z^n$.

7. Let $f(z) = \sum_{n=0}^{\infty} a_n z^n$ have radius of convergence 1, and suppose $f'(0) = a_1 \neq 0$. Show that there is an $r > 0$ such that f is one-to-one on $S(0, r)$.

8. Let $f(z) = \sum_{n=0}^{\infty} a_n z^n$ have radius of convergence $R > 0$, and suppose that $0 \leqslant r < R$. By writing

$$f(z) \cdot \overline{f(z)} = \sum_{n=0}^{\infty} \sum_{m=0}^{\infty} a_n \bar{a}_m r^{m+n} e^{i(n-m)\theta}$$

if $z = r e^{i\theta}$, $\theta \in [0, 2\pi]$, show that

$$\frac{1}{2\pi} \int_0^{2\pi} |f(r e^{i\theta})|^2 \, d\theta = \sum_{n=0}^{\infty} |a_n|^2 r^{2n}.$$

Deduce that if $|f(z)|$ has a local maximum at $z = 0$, then $a_n = 0$ if $n \geqslant 1$ (in other words, f is constant).

9. Let f be regular on $S(a, R)$ for some $a \in \mathbf{C}$, $R > 0$.
Show that if $|f|$ is constant on $S(a, R)$, then so is f.

10. Find, for each of the following curves, whether it is (a) closed, (b) smooth, (c) a path.
Give a sketch in each case, and find the lengths of those which are paths.

(i) $\gamma(t) = e^{3it}, \quad 0 \leqslant t \leqslant 1.$

(ii) $\gamma(0) = 0,\quad \gamma(t) = t + it^2 \sin\left(\dfrac{\pi}{2t}\right),\quad 0 < t \leqslant \frac{1}{2},$

$\gamma(t) = 1 - t,\quad \frac{1}{2} < t \leqslant 1.$

(iii) $\gamma(0) = 0,\quad \gamma(t) = \exp\left(\dfrac{i-1}{t}\right),\qquad 0 < t \leqslant 1.$

11. Let G be open in $\mathbf{C}$ and γ be a path with $\gamma^* \subset G$. Suppose f is regular on G, and $|f'| \leqslant M$ on γ^*. Show that $f \circ \gamma$ is also a path, and that

$$L(f \circ \gamma) \leqslant ML(\gamma).$$

12. Describe the components of the following open sets.
(i) $\{z\colon |z| < 1, \operatorname{Re} z \neq 0, \operatorname{Im} z > 0\}$,
(ii) $\{z\colon |z - 1| \neq 2, |z + 1| \neq 2\}$,
(iii) $\{z\colon z = x + iy, x > 0, y > 0, y \neq nx, n = 1, 2, 3, \ldots\}$.

13. Let $f(z) = e^z$ for $z \in \mathbf{C}$.
Evaluate the integrals of f and $|f|$ over the following paths,
(i) $[-1, 1]$, (ii) $[-i, i]$, (iii) $\gamma(t) = e^{i\pi t}$, $0 \leqslant t \leqslant 1$.

14. Let G be a connected open set, and f a regular function with $f' = 0$ on G.
Show that f is constant on G. (Hint: integrate f' along a polygonal path.)

15. Let ϕ be a continuous function from the interval $[-1, 1]$ to $\mathbf{C}$. For $z \in G = \mathbf{C} \setminus [-1, 1]$, define f by

$$f(z) = \int_{-1}^{1} \frac{\phi(t)\, dt}{t - z},$$

Show that f is regular on G, and find $f'(z)$ explicitly.

16. For an arbitrary set $E \subset \mathbf{C}$, consider the following property, which replaces (a) of 1.17:
(a′) Every decomposition of E into disjoint subsets E_1 and E_2, where $E_i = E \cap G_i$ ($i = 1, 2$) and G_1, G_2 are disjoint open subsets of $\mathbf{C}$, is trivial.
Show that this is equivalent to (c) of 1.17 (with E in place of G), but that (b) may not be equivalent if E is not open.
(a′) (or (c)) is the definition of connectedness for general subsets of $\mathbf{C}$.

2

Cauchy's Theorem

Our first aim is to prove Cauchy's theorem, which states that for a wide class of open sets and functions defined on them, the integral round a closed path in the open set is zero. The remainder of the chapter, and in a sense the whole of complex analysis, consists of the consequences of this result. In particular it is proved that a regular function has derivatives of all orders, and that the value of the function at a point inside a circle is determined by the values of the function on the circle. The classical theorems of Liouville and Morera are also proved.

§1. CAUCHY'S THEOREM FOR A STARRED DOMAIN

Definition 2.1. Let (a, b, c) be an ordered triple of complex numbers. We define the triangle

$$T(a, b, c) = \{z: z = \lambda a + \mu b + \nu c, \lambda, \mu, \nu \geqslant 0, \lambda + \mu + \nu = 1\}.$$

This is easily seen to be a closed bounded set, whose boundary ∂T is the union of the line segments $[a, b]$, $[b, c]$, $[c, a]$. This identifies ∂T with a closed path, and we define $\int_{\partial T} f = \int_{[a,b]} f + \int_{[b,c]} f + \int_{[c,a]} f$, for any f continuous on ∂T.

Notice that if the vertices a, b, c are interchanged cyclically, the value of $\int_{\partial T} f$ is unchanged, while for instance

$$\partial T(a, c, b) = -\partial T(a, b, c),$$

according to (iii) of definition 1.15.

Theorem 2.2. *Let G be an open set, $p \in G$, and suppose f is regular on*

$G \setminus \{p\}$, and continuous at p. Then for any triangle $T(a, b, c)$ contained in G, $\int_{\partial T} f(z)\,dz = 0$.

Proof. We suppose firstly that $p \notin T$, and proceed by a subdivision argument as follows.

We divide T into

$$
\begin{aligned}
T_1^{(1)} &= T(a, \tfrac{1}{2}(a + b), \tfrac{1}{2}(a + c)),\\
T_1^{(2)} &= T(b, \tfrac{1}{2}(b + c), \tfrac{1}{2}(b + a)),\\
T_1^{(3)} &= T(c, \tfrac{1}{2}(c + a), \tfrac{1}{2}(c + b)) \quad \text{and}\\
T_1^{(4)} &= T(\tfrac{1}{2}(a + b), \tfrac{1}{2}(b + c), \tfrac{1}{2}(c + a)).
\end{aligned}
$$

It is easily seen that

$$T = \bigcup_{j=1}^{4} T_1^{(j)} \quad \text{and that} \quad J = \int_{\partial T} f = \sum_{j=1}^{4} \int_{\partial T_1^{(j)}} f.$$

Denote by T_1 that triangle from $T_i^{(j)}, j = 1, 2, 3, 4$ which maximizes $|\int_{\partial T_1^{(j)}} f|$, and let $J_1 = \int_{\delta T_1} f$. Evidently $|J| \leqslant 4|J_1|$, and $L(\partial T_1) = \frac{1}{2}L(\partial T)$. We now repeat the subdivision with T_1 in place of T, thereby obtaining a sequence (T_n) of triangles, with the properties

$$
\begin{gathered}
T_n \subset T_{n-1} \subset \cdots \subset T \subset G,\\
|J| \leqslant 4|J_1| \leqslant 4^2|J_2| \leqslant \cdots \leqslant 4^n|J_n|,
\end{gathered}
$$

and $L(\partial T_n) = 2^{-n}L(\partial T)$. For each n, let z_n be any point of T_n—say the centroid for definiteness.

Then if $m > n$, $T_m \subset T_n$, and therefore $|z_m - z_n| < L(\partial T_n)$ and (z_n) is a Cauchy sequence in $\mathbf{C}$. Hence there is a point z_0 to which the sequence (z_n) converges, and since each T_n is closed, $z_0 \in \bigcap_{n=1}^{\infty} T_n$. In particular since $p \notin T$, $z_0 \neq p$, and f is regular at z_0.

We can write $f(z) = f(z_0) + A(z - z_0) + s(z)(z - z_0)$, where $A = f'(z_0)$ is constant, and $s(z) \to 0$ as $z \to z_0$.

Suppose $\varepsilon > 0$ is given. Choose $\delta > 0$ such that $|s(z)| < \varepsilon$ if $|z - z_0| < \delta$, and then n_0 such that $L(\partial T_n) < \delta$ for $n \geqslant n_0$.

Since $z_0 \in T_n$, it follows that if $z \in \partial T_n$, $|z - z_0| < L(\partial T_n)$ and $T_n \subset S(z_0, \delta)$, if $n \geqslant n_0$.

Now

$$
\begin{aligned}
J_n &= \int_{\partial T_n} f(z)\,dz\\
&= f(z_0) \int_{\partial T_n} dz + A \int_{\partial T_n} (z - z_0)\,dz + \int_{\partial T_n} s(z)(z - z_0)\,dz.
\end{aligned}
$$

Since

$$\frac{d}{dz}(z) = 1, \quad \text{and} \quad \frac{d}{dz}\left(\tfrac{1}{2}(z - z_0)^2\right) = z - z_0,$$

the first two terms in this summation are zero by (v) of theorem 1.20.

Hence

$$|J_n| = \left|\int_{\partial T_n} s(z)(z - z_0)\, dz\right| \leqslant L(\partial T_n) \cdot \varepsilon L(\partial T_n) \qquad \text{if} \quad n \geqslant n_0.$$

It follows that

$$|J| \leqslant 4^n |J_n| \leqslant 4^n \cdot [2^{-n}L(T)]^2 \varepsilon = \varepsilon L(T)^2.$$

Since ε is arbitrary, J must $= 0$.

We suppose now that $p \in T$, and consider the various possibilities which may arise.

If p is a vertex of T, say $p = a$, then we take points b_1 on $[a, b]$, and c_1 on $[a, c]$ and write

$$T = T(a, b_1, c_1) \cup T(b_1, b, c) \cup T(b_1, c, c_1) = T_1 \cup T_2 \cup T_3, \quad \text{say.}$$

Evidently $\int_{\partial T} f = \int_{\partial T_1} f + \int_{\partial T_2} f + \int_{\partial T_3} f$, and the second and third terms are zero since a is not a point of T_2 or T_3.

Now f is continuous at a, and so given $\varepsilon > 0$, there is a $\delta > 0$ such that $|f(z) - f(a)| < \varepsilon$ if $|z - a| < \delta$. We choose b_1, c_1 so that $|a - b_1|$ and $|a - c_1|$ are both $< \delta$. Then

$$\begin{aligned}\left|\int_{\delta T} f(z)\, dz\right| &= \left|\int_{\delta T_1} f(z)\, dz\right| \\ &\leqslant \left|\int_{\partial T_1} f(a)\, dz\right| + \left|\int_{\partial T_1} (f(z) - f(a))\, dz\right| \\ &\leqslant 0 + L(T_1) \cdot \varepsilon \leqslant L(T)\varepsilon.\end{aligned}$$

Since ε is arbitrary, it follows again that $\int_{\partial T} f = 0$. Finally if p is on a side, say $[a, b]$, of T we reduce it to the above case by writing $T = T(a, p, c) \cup T(p, b, c)$; while if p is a point of T not on any side, we write $T = T(a, b, p) \cup T(b, c, p) \cup T(c, a, p)$. This completes the proof in all cases.

We recall (from 1.16 (iii)) that a set $E \subset \mathbf{C}$ is said to be *starred* with respect to a point $a \in E$ if for all $z \in E$, the line segment $[a, z]$ is contained in E. The point a is called a star-centre for E.

Evidently any starred set is polygonally connected via the star centre, and hence if open, must be connected by 1.17.

Theorem 2.3. (Cauchy). *Let G be a starred open set in $\mathbf{C}$, $p \in G$, and suppose f is regular on $G \setminus \{p\}$, and continuous at p. Then for any closed path γ in G,*

$$\int_\gamma f(z)\,dz = 0.$$

Proof. After (v) of theorem 1.20, it is sufficient to prove that there is a function F which is regular on G, with $F'(z) = f(z)$ for all z in G.

Let a be a star-centre for G, and define $F(z) = \int_{[a,z]} f$ for each $z \in G$.

Since G is open, given $z \in G$, we can choose $r > 0$ such that $S(z, r) \subset G$. Then if $|h| < r$, $[z, z + h] \subset G$, and hence since G is starred with respect to a, $T(a, z, z + h) \subset G$.

If follows from theorem 2.2 that

$$0 = \int_{\partial T} f = \int_{[a,z]} f + \int_{[z,z+h]} f + \int_{[z+h,a]} f,$$

which may be rewritten as

$$F(z + h) = F(z) + \int_{[z,z+h]} f(u)\,du.$$

Since $\int_{[z,z+h]} f(z)\,du = hf(z)$, it follows that

$$\frac{1}{h}(F(z + h) - F(z)) - f(z) = \frac{1}{h}\int_{[z,z+h]} (f(u) - f(z))\,du.$$

Now f is assumed continuous at z, and so for any $\varepsilon > 0$, there is a $\delta > 0$ for which $|f(u) - f(z)| < \varepsilon$ if $|u - z| < \delta$.

Therefore if $|h| < \min(\delta, r)$,

$$\left|\frac{1}{h}(F(z + h) - F(z)) - f(z)\right| = \frac{1}{|h|}\left|\int_{[z,z+h]} (f(u) - f(z))\,du\right|$$

$$\leqslant \frac{1}{|h|} \cdot |h|\varepsilon = \varepsilon.$$

Hence F is regular, and $F'(z) = f(z)$ as required.

Notes 2.4. (i) It will be shown in corollary 2.9 that if f is regular on $G \setminus \{p\}$, and continuous at p, then in fact f must be regular at p also.

The exceptional point p is introduced in order to give a very simple proof of our next result.

(ii) The hypothesis that G be starred in theorem 2.3 is not the most precise known, but is adequate for all our applications. Some sort of condition on the shape of G is necessary, since for example the function $f(z) = 1/z$ is regular in $\mathbf{C} \setminus \{0\}$, while

$$\int_{C(0,1)} \frac{dz}{z} = 2\pi i$$

by lemma 1.22.

§2. INTEGRAL FORMULAE AND HIGHER DERIVATIVES

Theorem 2.5 (Cauchy integral formula). *Let G be an open set, and f be regular on G. Let $a \in G$ and $R > 0$ be such that $\bar{S}(a, R) \subset G$, and suppose $|w - a| < R$. Then*

$$f(w) = \frac{1}{2\pi i} \int_{C(a,r)} \frac{f(z)\,dz}{z - w}.$$

Proof. The sets $\bar{S}(a, R)$ and $\mathbf{C} \setminus G$ are disjoint and closed, and $\bar{S}(a, R)$ is compact. It follows from the property noted under (v) of definition 1.2, that there is a number $\delta > 0$ such that if $z_1 \in \bar{S}(a, R)$, $z_2 \in \mathbf{C} \setminus G$, then $|z_1 - z_2| \geqslant \delta$, that is $S(a, R + \delta) \subset G$.

Let $G_1 = S(a, R + \delta)$: G_1 is starred, and $C(a, R)$ is a closed path in G_1. Now define

$$h(z) = \frac{f(z) - f(w)}{z - w}$$

for $z \in G$, $z \neq w$, and $h(w) = f'(w)$.

Then h is regular on G_1 except at w, and is continuous at w, and we may apply theorem 2.3 to deduce that $\int_{C(a,r)} h(z)\,dz = 0$.

This may be rewritten

$$\int_{C(a,r)} \frac{f(z)\,dz}{z - w} = f(w) \int_{C(a,R)} \frac{dz}{z - w} = 2\pi i f(w)$$

by lemma 1.22.

Hence

$$f(w) = \frac{1}{2\pi i} \int_{C(a,R)} \frac{f(z)\,dz}{z - w} \quad \text{as required.}$$

Notes 2.6 This theorem together with the next one shows how the structure of a regular function is determined completely by its values on a circle within the domain of regularity. This contrasts strongly with the situation for a function of two real variables, which may have partial derivatives of all orders and vanish on the circle $x^2 + y^2 = 1$, and still not vanish identically: an example is given by the function

$$f(x, y) = \begin{cases} \exp\left(-\dfrac{1}{1 - x^2 - y^2}\right) & \text{for } x^2 + y^2 < 1. \\ 0 & \text{for } x^2 + y^2 \geqslant 1. \end{cases}$$

Theorem 2.7. *Let G be an open set, and f be regular in G. Then f has derivatives of all orders at all points of G. If $\bar{S}(a, R) \subset G$, and $|w - a| < R$, then for $k = 0, 1, 2, \ldots,$*

$$f^{(k)}(w) = \frac{k!}{2\pi i}\int_{C(a,R)} \frac{f(z)\,dz}{(z - w)^{k+1}}.$$

Proof. Since G is open, every point of G is contained in an open sphere in G, and it is therefore sufficient to prove the formula for the kth derivative. This we accomplish by induction, and theorem 2.5 provides us with the case $k = 0$.

Suppose therefore that for some value of $k \geqslant 0$, we have proved

$$f^{(k)}(w) = \frac{k!}{2\pi i}\int_{C(a,R)} \frac{f(z)\,dz}{(z - w)^{k+1}}$$

for $w \in S(a, R)$ and $\bar{S}(a, R) \subset G$.

The idea now is simply to differentiate the right-hand side of this expression with respect to w. This may either be done by appealing to theorem 1.23 on differentiation under the integral sign, or by an *ad hoc* argument as follows. Suppose that $w \in S(a, R)$, and that

$$|h| < \tfrac{1}{2}(R - |w - a|) = \tfrac{1}{2}\delta_0$$

say. Then certainly $w + h \in S(a, R)$. We have

$$^{(k)}(w + h) - f^{(k)}(w)$$

$$= \frac{k!}{2\pi i}\int_{C(a,R)} \left[\frac{1}{(z - w - h)^{k+1}} - \frac{1}{(z - w)^{k+1}}\right] f(z)\,dz,$$

and the expression in square brackets may be written

$$(k+1)\int_{[w,w+h]}\frac{du}{(z-u)^{k+2}}.$$

Hence

$$\frac{1}{h}\left(f^{(k)}(w+h)-f^{(k)}(w)\right)-\frac{(k+1)!}{2\pi i}\int_{C(a,R)}\frac{f(z)\,dz}{(z-w)^{k+2}}$$

$$=\frac{(k+1)!}{2\pi i}\int_{C(a,R)}\left[\frac{1}{h}\int_{[w,w+h]}\frac{du}{(z-u)^{k+2}}-\frac{1}{(z-w)^{k+2}}\right]f(z)\,dz.$$

The integrand is now

$$\frac{f(z)}{h}\int_{[w,w+h]}\left(\frac{1}{(z-u)^{k+2}}-\frac{1}{(z-w)^{k+2}}\right)du$$

$$=\frac{(k+2)f(z)}{h}\int_{[w,w+h]}\left(\int_{[w,u]}\frac{dv}{(z-v)^{k+3}}\right)du.$$

Now $v\in[w,u]\subset[w,w+h]$, and

$$|z-v|\geqslant R-|v-a|\geqslant R-|w-a|-|h|\geqslant\tfrac{1}{2}\delta_0.$$

Hence the integrand is in modulus

$$\leqslant\frac{(k+2)}{|h|}\sup_{z\in C(a,R)}|f(z)|\cdot|h|\cdot|h|(\tfrac{1}{2}\delta_0)^{-k-3},$$

which tends to zero with h.

Hence $f^{(k)}$ is differentiable at w, and its derivative is given by

$$\frac{(k+1)!}{2\pi i}\int_{C(a,R)}\frac{f(z)\,dz}{(z-w)^{k+1}}\quad\text{as required.}$$

§3. MORERA'S AND LIOUVILLE'S THEOREMS

The following theorem is a kind of converse to Cauchy's theorem in that it asserts that if the integral of a continuous function around every triangular path is an open set is zero, then the function must be regular.

Theorem 2.8 (Morera). *Let G be an open set, f be continuous on G, and suppose that for every triangle $T\subset G$, $\int_{\partial T}f(z)\,dz=0$. Then f is regular on G.*

Proof. Let a be any point of G, and choose $R > 0$ so that $S(a, R) \subset G$.

For $z \in S(a, R)$, let $F(z) = \int_{[a,z]} f(u)\, du$. If $z + h$ is also in $S(a, R)$, the hypothesis on $T(a, z, z + h)$ shows that

$$F(z + h) = F(z) + \int_{[z,z+h]} f(u)\, du.$$

It follows as in theorem 2.2 that since f is continuous, $F'(z) = f(z)$ at all points of $S(a, R)$. Hence F is regular, and by theorem 2.7 F has derivatives of all orders. In particular, $F' = f$ is differentiable in $S(a, R)$, that is f is regular in $S(a, R)$. But a was any point of G, and so f is regular throughout G.

Corollary 2.9. *Let a function f satisfy the hypotheses of theorem 2.2 on an open set G. Then f is regular in G.*

Proof. The conclusion of theorem 2.2 is that $\int_{\partial T} f(z)\, dz = 0$ for all triangles $T \subset G$. Hence f is regular in G by Morera's theorem.

Lemma 2.10 (Cauchy's inequalities). *Let f be regular in an open set G, $\bar{S}(a, R) \subset G$ and $|f(z)| \leqslant M$ for $z \in C(a, R)$. Then for $k = 0, 1, 2, \ldots$, and $w \in S(a, R)$,*

$$|f^{(k)}(w)| \leqslant \frac{k!MR}{(R - |w - a|)^{k+1}}.$$

In particular $|f^{(k)}(a)| \leqslant k!M/R^k$.

Proof. From theorem 2.7 we have

$$f^{(k)}(w) = \frac{k!}{2\pi i} \int_{C(a,R)} \frac{f(z)\, dz}{(z - w)^{k+1}}.$$

If $z \in C(a, R)$, $w \in S(a, R)$, then $|z - w| \geqslant R - |w - a|$.
Hence

$$|f^{(k)}(w)| \leqslant \frac{k!}{2\pi} \frac{2\pi RM}{(R - |w - a|)^{k+1}} = \frac{k!MR}{(R - |w - a|)^{k+1}}$$

as required.

Theorem 2.11 (Liouville). *Let f be regular and bounded on* $\mathbf{C}$. *Then f must be constant.*

Proof. Let $w \in \mathbf{C}$, and $R > 0$ be arbitrary. Suppose $|f(z)| \leqslant M$ for all z. Then

$$|f'(w)| \leqslant \frac{1!M}{R}$$

on applying 2.10 to f on $S(w, R)$. Since M is independent of R, and R is unrestricted, $f'(w)$ must $=0$, for all $w \in \mathbf{C}$. It follows that for all $w \in \mathbf{C}$,

$$f(w) = f(0) + \int_{[0,w]} f'(z)\, dz = f(0),$$

and f is constant.

Theorem 2.12 (Fundamental theorem of algebra). *Let $P(z)$ be a non-constant polynomial, that is $P(z) = a_0 + a_1 z + \cdots + a_n z^n$ for some $n \geqslant 1$, and $a_n \neq 0$. Then for some $z_0 \in \mathbf{C}$, $P(z_0) = 0$.*

Proof. Since $a_n \neq 0$,

$$\begin{aligned}|P(z)| &\geqslant |a_n||z|^n - |a_{n-1}||z|^{n-1} - \cdots - |a_1||z| - |a_0| \\ &= |a_n||z|^n\left(1 - \frac{|a_{n-1}|}{|a_n||z|} - \cdots - \frac{|a_1|}{|a_n||z|^{n-1}} - \frac{|a_0|}{|a_n||z|^n}\right).\end{aligned}$$

If we choose $|z|$ to be >1 and $>2n|a_j|/|a_n|$ for each $j = 0,1,\ldots, n-1$, then

$$\frac{|a_j|}{|a_n||z|^{n-j}} < \frac{1}{2n}, \quad \text{and hence} \quad |P(z)| > \tfrac{1}{2}|a_n||z|^n,$$

that is $|P(z)| \to \infty$ as $|z| \to \infty$.

It follows that if $P(z)$ is a non-constant polynomial which is never zero, then $1/[P(z)]$ is a regular function for all z, and tends to zero as $|z| \to \infty$.

Hence $1/[P(z)]$ is bounded, and must be constant by Liouville's theorem. Since $1/[P(z)] \to 0$ at infinity, we must have $1/[P(z)] = 0$ for all z, which is clearly impossible.

Notes 2.13. (i) Liouville's theorem has many generalizations in which a condition on the behaviour of a regular function for large z, leads to very precise results about the nature of the function.

(ii) It may seem strange to a beginning student to find a proof of the fundamental theorem of algebra in this context. However, some analytic ideas are essential to the proof of this result—at the very least one needs the properties of the exponential function listed in theorem 1.8.

EXERCISES FOR CHAPTER 2

1. To which of the following situations is Cauchy's theorem applicable?

(i) $G = \mathbf{C}\setminus\{i\};\ f(z) = (z - i)^{-1}, z \in G;$
$\gamma(t) = 1 + e^{2\pi it},\quad 0 \leqslant t \leqslant 1.$

(ii) $G = \mathbf{C}\setminus\{1\};\ f(z) = (z - 1)^{-1}, z \in G;$
$\gamma(t) = 1 + e^{2\pi it},\quad 0 \leqslant t \leqslant 1.$

(iii) $G = \mathbf{C};\ f(z) = \sin z/z, z \neq 0, f(0) = 0;$
$\gamma(t) = \frac{1}{2}e^{2\pi it},\quad 0 \leqslant t \leqslant 1.$

(iv) $G = \mathbf{C};\ f(z) = \sin z/z^2, z \neq 0, f(0) = 0;$
$\gamma(t) = 2e^{2\pi it},\quad 0 \leqslant t \leqslant 1.$

(v) $G = \mathbf{C};\ f(z) = z \sin z, z \in G;\quad \gamma(t) = 1 + it,\quad 0 \leqslant t \leqslant 1.$

(vi) $G = \mathbf{C}\setminus\mathbf{Z};\ f(z) = (e^{2\pi iz} - 1)^{-1}, z \in G;$
$\gamma(t) = \frac{1}{2} + \frac{1}{2}i + e^{2\pi it},\quad 0 \leqslant t \leqslant 1.$

2. Let G be open in $\mathbf{C}$, and f be continuous on $G \times [0, 1]$.

Suppose for each $t \in [0, 1]$, $f(z, t)$ is a regular function of $z \in G$. Use Morera's theorem to show that $F(z) = \int_0^1 f(z, t)\, dt$ is regular on G (cf. 1.23).

3. Use the Cauchy integral formulae to evaluate

(i) $\int_{C(0,R)} e^z z^{-k-1}\, dz,\quad k = 0, 1, 2, \ldots,$

(ii) $\dfrac{1}{2\pi i}\displaystyle\int_{C(0,1)} z(z - a)^{-1}\, dz,\quad |a| < 1,$

(iii) $\dfrac{1}{2\pi i}\displaystyle\int_{C(1,2)} \sin z\{z(z + 2)\}^{-1}\, dz,$

(iv) $\dfrac{1}{2\pi i}\displaystyle\int_{C(i,2)} (\sin z/z)^2\, dz,$

(v) $\int_{C(-i,2)} (z - 1)(z - 2)^{-1}(z + 1)^{-2}\, dz.$

4. Let f be regular on $S(0, R)$, and suppose $|a| < r < R$.
Prove that

$$f(a) = \frac{1}{2\pi i}\int_{C(0,r)} \frac{(r^2 - |a|^2)f(z)}{(r^2 - z\bar{a})(z - a)}\, dz.$$

Deduce that if $a = |a|\, e^{i\theta}$ $(\theta \in \mathbf{R})$, then

$$f(a) = \frac{1}{2\pi}\int_0^{2\pi} \frac{(r^2 - |a|^2)f(r\, e^{i\phi})\, d\phi}{r^2 - 2|a|r\cos(\theta - \phi) + |a|^2}.$$

5. Let f be regular on $\mathbf{C}$, and satisfy an inequality of the form $|f(z)| \leqslant A + B|z|^k$, where $k = 0, 1, 2, \ldots$, and A, B are positive constants.

Use 2.10 to show that $f^{(k+1)}(z) = 0$ for all $z \in \mathbf{C}$, and hence that f must be a polynomial of degree $\leqslant k$.

6. Let f be continuous on $\mathbf{C}$, and define

$$M(r) = \sup\{|f(z)| : |z| = r\},$$
$$m(r) = \inf\{|f(z)| : |z| = r\}.$$

Calculate $M(r)$ and $m(r)$ for the following functions:

(i) $f(z) = z^n$ $(n = 0, 1, 2, \ldots)$, $z \in \mathbf{C}$,
(ii) $f(z) = e^z$, $z \in \mathbf{C}$,
(iii) $f(z) = z^2 - 1$, $z \in \mathbf{C}$,
(iv) $f(z) = \cos z$, $z \in \mathbf{C}$.

The functions M and m are important in describing the behaviour of f as $r \to \infty$ (compare exercise 5 above, whose hypothesis simply says $M(r) \leqslant A + Br^k$).

7. Let f be regular in $S(0, 1)$, and suppose $|f(z)| \leqslant (1 - |z|)^{-1}$ for all $z \in S(0, 1)$.

Use the integral formula

$$f^{(n)}(0) = \frac{n!}{2\pi i}\int_{C(0,r)} f(z)z^{-n-1}\, dz, \qquad 0 < r < 1,$$

to estimate $|f^{(n)}(0)|$.

Which value of r gives the sharpest estimate for $f^{(n)}(0)$?

8. Suppose f satisfies the same conditions as in exercise 7. Let $a \in S(0, 1)$, and use the integral formula (with $|a| < r < 1$) to give an estimate for $|f^{(n)}(a)|$. Can you still find a most favourable value of r?

9. Let f be regular on $S(0, R)$ with $R > 1$, and suppose $f(x)$ is real if $-1 \leqslant x \leqslant 1$.

Prove that $\int_{-1}^{1} f^2(x)\,dx \leqslant \frac{1}{2}\int_0^{2\pi} |f(e^{i\theta})|^2\,d\theta$.
[Hint: Apply Cauchy's theorem to the function f^2 and the two closed paths given by the boundaries of the upper and lower semicircles in the unit disc $\bar{S}(0, 1)$.]
By taking

$$f(z) = \sum_{j=0}^{n} a_j z^j \qquad (a_j \in \mathbf{R}, \quad j = 0, 1, 2, \ldots, n),$$

deduce that

$$\sum_{j=0}^{n} \sum_{k=0}^{n} \frac{a_j a_k}{j + k + 1} \leqslant \pi \sum_{j=0}^{n} a_j^2.$$

3

Local Properties of Regular Functions

In this chapter we prove results about regular functions which are local in character—that is they describe the behaviour of a function in an open sphere $S(a, R)$ in which it is regular. We also investigate what happens to a function which is regular in a set of the form $S(a, R) \setminus \{a\}$, but not regular at a.

We begin with Taylor's theorem which shows that every function regular in some $S(a, R)$ can be represented by a power series there.

§1. TAYLOR'S THEOREM

Theorem 3.1 (Taylor). *Let f be regular in an open set G, and $S(a, R) \subset G$ for some $a \in G$, $R > 0$.*

Then for all $z \in S(a, R)$, $f(z) = \sum_{n=0}^{\infty} a_n(z - a)^n$, where the series is absolutely convergent in $S(a, R)$, and the coefficients are given by

$$a_n = \frac{1}{n!} f^{(n)}(a), \qquad n = 0, 1, 2, \ldots.$$

Proof. Let f, $S(a, R)$ be as in the statement of the theorem, and for any $w \in S(a, R)$ choose a value of r with $|w - a| < r < R$. According to theorem 2.7,

$$f(w) = \frac{1}{2\pi i} \int_{C(a,r)} \frac{f(z)\, dz}{z - w}.$$

Since $|w - a| < r = |z - a|$ for $z \in C(a, r)$, we can write

$$\frac{1}{z - w} = \frac{1}{(z - a) - (w - a)}$$

$$= \frac{1}{z - a}\left(1 - \frac{w - a}{z - a}\right)^{-1} = \sum_{n=0}^{\infty} \frac{(w - a)^n}{(z - a)^{n+1}},$$

where the series is uniformly and absolutely convergent for $z \in C(a, r)$ by Weierstrass' M-test. Since f is bounded on $C(a, r)$, the uniform convergence with respect to z is not affected on multiplying through by $f(z)$, and hence to evaluate

$$\frac{1}{2\pi i}\int_{C(a,R)} \frac{f(z)\, dz}{z - w}$$

we can integrate term by term. We obtain

$$f(w) = \sum_{n=0}^{\infty} (w - a)^n \frac{1}{2\pi i}\int_{C(a,r)} \frac{f(z)\, dz}{(z - a)^{n+1}} = \sum_{n=0}^{\infty} a_n(w - a)^n,$$

and the coefficient

$$a_n = \frac{1}{2\pi i}\int_{C(a,r)} \frac{f(z)\, dz}{(z - a)^{n+1}} = \frac{f^{(n)}(a)}{n!}$$

by theorem 2.7.

Note 3.2. The series $\sum_{n=0}^{\infty} (f^{(n)}(a)/n!)(z - a)^n$ is called the Taylor series for f about the point a. Theorem 3.1 guarantees that this series will converge to the function f on any open sphere which is contained in the set of regularity.

The student will have learned from courses in real analysis that a function of one real variable need not have a convergent Taylor series, and that even if the series converges its sum may not equal the function itself. In the complex case the situation is a little better, since the Taylor series does converge to f in some open sphere. However, we do not assert that the radius R with $S(a, R) \subset G$ is the radius of convergence of the Taylor series (which we will denote by R_0); only that $R \leqslant R_0$. If $R_0 > R$ the Taylor series will converge in $S(a, R_0)$, to a function f_0 which is regular in $S(a, R_0)$ by (ii) of theorem 1.13.

Taylor's theorem simply asserts that $f = f_0$ in $S(a, R)$. The following examples illustrate some of the possibilities.

Examples 3.3. (i) The function $f(z) = 1/(z - i)$ is regular on $\mathbf{C} \setminus \{i\}$. If $z = x$ is any point on the real axis, we may expand f about x by the binomial theorem to obtain

$$f(z) = \frac{1}{(z - x) - (i - x)} = - \sum_{n=0}^{\infty} \frac{(z - x)^n}{(i - x)^{n+1}},$$

the expansion being valid for $|z - x| < |i - x|$. But $|i - x| = \sqrt{(1 + x^2)}$ which is the radius of the largest sphere centred at x, contained in $\mathbf{C} \setminus \{i\}$. In this case the radius of convergence of the Taylor series coincides with the radius of the largest sphere contained in the domain of regularity.

(ii) Let $f(z) = \int_0^1 (z\,dt)/(1 + tz)$. The integral is well defined provided $tz \neq -1$ for $0 \leqslant t \leqslant 1$, i.e. provided z is not real and $\leqslant -1$. We suppose then that

$$z \in D = \mathbf{C} \setminus \{x + iy; \quad y = 0, \quad x \leqslant -1\}.$$

It is tempting to 'evaluate' the integral defining f by logarithms—however, the reader is reminded that we have no definition of $\log(x)$ except in the case when x is real and positive.

It follows by differentiating the integral with respect to z, that f is regular on D, with

$$f'(z) = \int_0^1 \left(\frac{1}{1 + tz} - \frac{zt}{(1 + tz)^2} \right) dt = \int_0^1 \frac{1}{(1 + tz)^2}\, dt.$$

This integral, which is again well defined on D, may be evaluated to give

$$\frac{1}{z} \left(- \frac{1}{1 + tz} \right) \Bigg|_{t=0}^{t=1} = \frac{1}{z} \left(1 - \frac{1}{1 + z} \right) = \frac{1}{1 + z}$$

if $z \neq 0$, while if $z = 0$, the value is trivially $=1$.

Hence $f'(z) = 1/(1 + z)$ for all $z \in D$, and by induction

$$f^{(n)}(z) = \frac{(-1)^{n-1}(n - 1)!}{(1 + z)^n}, \qquad n = 1, 2, 3, \ldots.$$

If a is any point on the circle $C(-1, 1)$, $(a \neq -2)$, then $|1 + a| = 1$, and the Taylor series about a,

$$\sum_{n=0}^{\infty} \frac{f^{(n)}(a)}{n!} (z - a)^n = f(a) + \sum_{n=1}^{\infty} \frac{(-1)^{n-1}}{n(1 + a)^n} (z - a)^n$$

has radius of convergence 1, the distance from a to the point -1.

However, if say $z = -1 + \omega$, where $\omega = -\frac{1}{2} + [\sqrt{3}/2]i$, the largest R for which $S(-1 + \omega, R) \subset D$ is $\sqrt{3}/2$ which is less than 1.

The explanation of this difficulty lies in the fact that in the portion of $S(-1 + \omega, 1)$ which lies below the negative real axis, the Taylor series about $-1 + \omega$ is convergent to a function not equal to $f(z)$ (in fact differing from it by $2\pi i$).

(iii) The problems in the real case are amply illustrated by the following two well-known examples.

(1) The function $f(x) = \exp(-x^{-2})$, $x \neq 0$, $f(0) = 0$ is infinitely differentiable at $x = 0$, with $f^{(n)}(0) = 0$, all n. Hence the Taylor series $\sum_{n=0}^{\infty} [f^{(n)}(0)/n!]x^n$ is convergent—in fact vanishes identically—for all x, but its sum is zero, not $f(x)$.

(2) The function $f(x) = \int_0^{\infty} [e^{-t}/(1 + x^2 t)]\, dt$ is defined and infinitely differentiable for all real x; its Taylor series about $x = 0$ is $\sum_{n=0}^{\infty} (-1)^n n! x^{2n}$ which diverges for all $x \neq 0$.

Needless to say, in neither of these cases can the definition of f be extended to complex z in such a way as to make it regular at $z = 0$.

We now prove two very important consequences of Taylor's theorem. The first of these is the uniqueness theorem for regular functions, in which we begin to make use of the notion of connectedness developed in Chapter 1. The second tells us that the radius of convergence of a power series, about which Taylor's theorem gives us no precise information, is determined by the existence of singularities; that is f cannot be defined in such a way as to be regular at all points of the circle of convergence.

Theorem 3.4. *Let f_1, f_2 be regular on open sets G_1, G_2, and suppose $G_1 \cap G_2$ is connected. Let $A \subset G_1 \cap G_2$ have a limit point in $G_1 \cap G_2$ and suppose $f_1 = f_2$ at all points of A. Then $f_1 = f_2$ at all points of $G_1 \cap G_2$.*

Proof. We write D for the connected open set $G_1 \cap G_2$, and $f = f_1 - f_2$ on D. Then we have $f = 0$ on A, and we require $f = 0$ on D.

Since A has a limit point in D, we may take a sequence (a_n) of points of A, $a_n \to a \in D$ as $n \to \infty$, and $a_n \neq a$, all n.

f is regular at $a \in D$, and so may be represented by its Taylor series about $z = a$,

$$f(z) = \sum_{n=0}^{\infty} c_n(z-a)^n \quad \text{for} \quad z \in S(a, R) \subset D.$$

Now f is continuous at a, $f(a_n) = 0$ all n, and $a_n \to a$ as $n \to \infty$, so $f(a)$ must $=0$, and hence $c_0 = 0$.

Therefore

$$f_1(z) = \frac{f(z)}{z-a} = \sum_{n=1}^{\infty} a_n(z-a)^{n-1},$$

is also continuous at a, and $f_1(a_n) = 0$ since $a_n \neq a$. Hence $f_1(a) = 0$, and $c_1 = 0$. We now consider successively

$$f_n(z) = \frac{f(z)}{(z-a)^n}, \qquad n = 1, 2, \ldots$$

to prove that each coefficient c_n must $=0$, and so $f(z) = 0$ in $S(a, R)$.

To complete the proof we have to show that f vanishes on the whole of D, and for this we need to use the connectedness of D.

Let b be any other point of D. By theorem 1.17 there is a polygonal path γ in D, with $\gamma(0) = a$, $\gamma(1) = b$.

Let $t_0 = \sup\{t: f(\gamma(u)) = 0 \text{ for } 0 \leqslant u \leqslant t\}$.

Since γ is continuous at 0, and $f = 0$ in $S(a, R)$, it follows that $t_0 > 0$. From the definition of t_0, there is a sequence (t_n) of points with $t_n < t_0$, $t_n \to t_0$ as $n \to \infty$, and $f(\gamma(t_n)) = 0$ for each n. We may therefore apply the argument above to deduce the existence of a sphere $S(\gamma(t_0), R_0) \subset D$ on which $f = 0$.

But then either $t_0 = 1$, or by the continuity of γ at t_0, $f(\gamma(t)) = 0$ on some interval $[t_0, t_0 + \delta]$, contrary to the definition of t_0.

Hence t_0 must $=1$, and $f(\gamma(1)) = 0$, or $f(b) = 0$ as required.

Theorem 3.5. *Suppose $f(z) = \sum_{n=0}^{\infty} a_n(z-a)^n$, where the series has finite radius of convergence $R > 0$. Then it is not possible to define f to be regular at all points of $C(a, R)$.*

Proof. We suppose the contrary, that is that the definition of f may be extended to make f regular at each point of $C(a, R)$. It follows from definition 1.12 that for each $z \in C(a, R)$, there is a positive radius $r_z > 0$, and a function f_z which is regular on $S(z, r_z)$ and $f = f_z$ on $S(z, r_z) \cap S(a, R)$.

Suppose then that $z_1, z_2 \in C(a, R)$ are such that

$$D = S(z_1, r_{z_1}) \cap S(z_2, r_{z_2}) \neq \varnothing .$$

Then $f_{z_1} = f_{z_2} = f$ on $D \cap S(a, R)$. Hence by 3.4 above $f_{z_1} = f_{z_2}$ on D, that is the functions f_{z_1}, f_{z_2} must agree at any point where both are defined. It follows that if

$$G = S(a, R) \cup \Big\{ \bigcup_{z \in C(a,R)} S(z, r_z) \Big\}, \quad \text{then} \quad F = \begin{cases} f & \text{on} \quad S(a, R) \\ f_z & \text{on} \quad S(z, r_z) \end{cases}$$

is a well-defined function which is regular on G, and has the same Taylor expansion as f about a.

By (v) (B) of 1.2, there is $\delta > 0$ such that $S(a, R + \delta) \subset G$, and it now follows from Taylor's theorem that the radius of convergence of the Taylor expansion of f is at least $R + \delta$, which contradicts our hypothesis.

Notes 3.6. (i) The Taylor coefficients a_n are evidently unique. For if $f(z) = \sum_0^\infty b_n(z - a)^n$ in some $S(a, r)$, then by 1.14,

$$b_n = \frac{1}{n!} f^{(n)}(a) = a_n.$$

(ii) The series $\sum_{n=1}^\infty (1/n^2)z^n$ has radius of convergence 1, and is absolutely and uniformly convergent on $\bar{S}(0, 1)$. Theorem 3.5 tells us that not all points of $C(0, 1)$ can be points of regularity, but does not help identifying the singularities (see definition 1.12). The following theorem of Pringsheim, which is not needed in the sequel, gives some information: Let

$$f(z) = \sum_0^\infty a_n(z - a)^n$$

have radius of convergence $R > 0$. Then if each $a_n \geqslant 0$, the point $a + R$ is a singularity of f.

For example $z = 1$ is a singularity of $\sum_1^\infty (1/n^2)z^n$.

The function given by $f(z) = (1 - z)^{-1}$, $(z \neq 1)$ has a Taylor series $\sum_0^\infty z^n$ about $z = 0$, whose radius of convergence is also 1. The series diverges at all points of $C(0, 1)$ although f is regular everywhere except at $z = 1$.

These examples show that no conclusion about the regularity of the sum function at points on the circle of convergence can be drawn from the convergence or divergence of the Taylor series there.

§2. LAURENT EXPANSIONS

We now consider functions which are regular on a domain of the form $\{z: r < |z - a| < R\}$ where $0 \leqslant r < R \leqslant \infty$. Such a set is called an annulus, and will be denoted $A(a; r, R)$. We shall be particularly interested in the case where $r = 0$, when the annulus reduces to $S(a, R) \setminus \{a\}$.

Theorem 3.7 (Laurent). *Let f be regular on an open set G, and $A(a; r, R) \subset G$.*

Then $f(z) = \sum_{-\infty}^{\infty} a_n(z - a)^n$, where the series converges absolutely for all $z \in A(a; r, R)$ and the coefficients are given by

$$a_n = \frac{1}{2\pi i}\int_{C(a,\rho)} \frac{f(z)\, dz}{(z - a)^{n+1}} \quad \text{for all} \quad n = 0, \pm 1, \pm 2, \ldots,$$

and any ρ with $r < \rho < R$.

Before beginning the proof we prove a technical result which simplifies the later argument.

Lemma 3.8. *Let h be regular on $A(a; r, R)$. Then if $r < \rho < R$, $\int_{C(a,\rho)} h(z)\, dz$ is independent of ρ.*

Proof. Let $g(z) = (z - a)h(z)$. g is also regular in A, and in particular it has a continuous derivative on A.

For $r < \rho < R$, define

$$\phi(\rho) = \int_{C(a,\rho)} h(z)\, dz = \int_{C(a,\rho)} \frac{g(z)}{z - a}\, dz = \int_0^{2\pi} g(a + \rho\, e^{i\theta}) i\, d\theta.$$

Then

$$\phi'(\rho) = \int_0^{2\pi} g'(a + \rho\, e^{i\theta}) i\, e^{i\theta}\, d\theta = \frac{1}{\rho} g(a + \rho\, e^{i\theta})\Big|_{\theta=0}^{2\pi} = 0$$

for all ρ with $r < \rho < R$. It follows that $\phi(\rho)$ is constant, as required.

Proof of theorem 3.7. Let $w \in A(a; r, R)$ and choose r_1, R_1 with $r < r_1 < |w - a| < R_1 < R$. Define

$$h(z) = \frac{f(z) - f(w)}{z - w},$$

if $z \in A$, $z \neq w$, and $h(w) = f'(w)$. By corollary 2.9, h is regular in A, and it follows from lemma 3.8 that

$$\int_{C(a,r_1)} h(z)\,dz = \int_{C(a,R_1)} h(z)\,dz.$$

This equation may be rewritten as

$$\int_{C(a,r_1)} \frac{f(z) - f(w)}{z - w}\,dz = \int_{C(a,R_1)} \frac{f(z) - f(w)}{z - w}\,dz,$$

and we know from 1.22 that

$$\int_{C(a,r_1)} \frac{dz}{z - w} = 0, \quad \text{and} \quad \int_{C(a,R_1)} \frac{dz}{z - w} = 2\pi i.$$

Hence

$$\int_{C(a,R_1)} \frac{f(z)\,dz}{z - w} - \int_{C(a,r_1)} \frac{f(z)\,dz}{z - w} = f(w)(2\pi i - 0),$$

or

$$f(w) = \frac{1}{2\pi i}\int_{C(a,R_1)} \frac{f(z)\,dz}{z - w} - \frac{1}{2\pi i}\int_{C(a,r_1)} \frac{f(z)\,dz}{z - w}$$

$$= f_1(w) + f_2(w), \quad \text{say.}$$

We now expand f_1 and f_2 in a series of powers of $(w - a)$ as was done in the proof of Taylor's theorem.

If $z \in C(a, R_1)$, then $|z - a| = R_1 > |w - a|$, and we write

$$\frac{1}{z - w} = \frac{1}{z - a}\left(1 - \frac{w - a}{z - a}\right)^{-1} = \sum_{n=0}^{\infty} \frac{(w - a)^n}{(z - a)^{n+1}},$$

where the series is uniformly convergent with respect to z. Hence

$$f_1(w) = \sum_{n=0}^{\infty} a_n(w - a)^n, \quad \text{where} \quad a_n = \frac{1}{2\pi i}\int_{C(a,R_1)} \frac{f(z)\,dz}{(z - a)^{n+1}}.$$

By 3.8 the path of integration may be replaced by $C(a, \rho)$, for any $r < \rho < R$.

Similarly if $z \in C(a, r_1)$, then $|z - a| = r_1 < |w - a|$, and we write

$$\frac{1}{z - w} = -\frac{1}{w - a}\left(1 - \frac{z - a}{w - a}\right)^{-1} = -\sum_{k=0}^{\infty} \frac{(z - a)^k}{(w - a)^{k+1}},$$

and again the series converges uniformly in z. Hence

$$f_2(w) = -\frac{1}{2\pi i}\int_{C(a,r_1)} \frac{f(z)\,dz}{z - w} = \sum_{k=0}^{\infty} b_k(w - a)^{-k-1},$$

where

$$b_k = \frac{1}{2\pi i}\int_{C(a,r_1)} (z - a)^k f(z)\,dz.$$

On writing $n = -k - 1$, we obtain

$$f_2(w) = \sum_{n=-\infty}^{-1} a_n(w - a)^n$$

with

$$a_n = b_{-n-1} = \frac{1}{2\pi i}\int_{C(a,r_1)} \frac{f(z)\,dz}{(z - a)^{n+1}} = \frac{1}{2\pi i}\int_{C(a,\rho)} \frac{f(z)\,dz}{(z - a)^{n+1}}$$

by 3.8.

Notes 3.9. (i) The series $\sum_{-\infty}^{\infty} a_n(z - a)^n$ is called the Laurent series of f in the annulus $A(a; r, R)$. $f_1(z) = \sum_{n=0}^{\infty} a_n(z - a)^n$ defines a regular function in $S(a, R)$, while $f_2(z) = \sum_{-\infty}^{-1} a_n(z - a)^n$ defines a regular function in the complement of $\bar{S}(a, r)$. In the case $r = 0$, f_2 is called the principal part of f at a. The series representation of f_2 shows that it is regular on $\mathbf{C} \setminus \{a\}$, and the series converges uniformly on $\mathbf{C} \setminus \bar{S}(a, r)$, $r > 0$.

If f is actually regular at a, then $f_2 = 0$ and the Laurent series reduces to the Taylor series about a. Notice that except in this particular case, we do not have an interpretation of the coefficients a_n in terms of derivatives of f at a.

(ii) The coefficients a_n are uniquely determined, for suppose that a series $\sum_{-\infty}^{\infty} b_n(z - a)^n$ also converged to f in $A(a; r, R)$. A standard power series argument (applied separately to the series $\sum_{-\infty}^{-1}$ and $\sum_{0}^{\infty}$ shows the convergence must be absolute and uniform on any $C(a, \rho)$, $r < \rho < R$.

Since for any integer k, and $z \in C(a, \rho)$,

$$\sum_{-\infty}^{\infty} a_n(z-a)^{n-k-1} = \sum_{-\infty}^{\infty} b_n(z-a)^{n-k-1} \quad \text{and} \quad \int_{C(a,\rho)} (z-a)^p \, dz = 0$$

if $p \neq -1$ or $= 2\pi i$ if $p = -1$, we obtain $a_k = b_k$ on integrating term by term.

Example 3.10. The function given by $f(z) = 1/[(z-1)(z-2)]$ has three Laurent expansions about $z = 0$.

In $S(0, 1)$, $f(z) = \sum_{n=0}^{\infty} (1 - 2^{-n-1})z^n$, which is simply the Taylor series about the origin. In $A(0; 1, 2)$,

$$f(z) = -\sum_{n=-\infty}^{-1} z^n - \sum_{n=0}^{\infty} 2^{-n-1}z^n,$$

while in $A(0; 2, \infty)(= \mathbf{C} \setminus \bar{S}(0, 2))$, $f(z) = \sum_{-\infty}^{-1} (2^{-n-1} - 1)z^n$.

Each of these expansions may be obtained by writing

$$f(z) = \frac{1}{z-2} - \frac{1}{z-1}.$$

and using the Binomial expansion.

The general situation concerning the uniform convergence of sequences of regular functions is described in the following theorem, which depends only on the results of Chapter 2.

Theorem 3.11. *Let (f_n) be a sequence of functions each of which is regular on an open set G. Suppose there is a function f on G having the property that for each compact set $K \subset G$, $f_n \to f$ uniformly on K as $n \to \infty$. Then*

(i) *f is regular on G, and*
(ii) *for each compact $K \subset G$, and integer $k \geqslant 1$, $f_n^{(k)} \to f^{(k)}$ uniformly on G.*

Proof. Since each point of G is a compact set, we certainly have $f_n \to f$ at each point of G. Also f is continuous, being the uniform limit of continuous functions. Now let T be any triangle contained in G. Theorem 2.2 shows that $\int_{\partial T} f_n = 0$ for each n.

But ∂T is compact, and hence $f_n \to f$ uniformly on ∂T. It follows that

$$\int_{\partial T} f = \lim_{n\to\infty} \int_{\partial T} f_n = 0.$$

Theorem 2.8 now tells us that f is regular.

Suppose that K is a compact subset of G.

For each $a \in K$, take $r > 0$ such that $\bar{S}(a, r) \subset G$, and let

$$N(a) = S(a, \tfrac{1}{2}r).$$

If $w \in N(a)$, $k \geqslant 1$,

$$f_n^{(k)}(w) = \frac{k!}{2\pi i}\int_{C(a,r)} \frac{f_n(z)\,dz}{(z-w)^{k+1}}.$$

Also

$$\frac{1}{|z-w|^{k+1}} \leqslant \frac{1}{(\frac{1}{2}r)^{k+1}} \quad \text{if} \quad z \in C(a, r).$$

It follows that as $n \to \infty$,

$$\frac{f_n(z)}{(z-w)^{k+1}} \to \frac{f(z)}{(z-w)^{k+1}}$$

uniformly for $z \in C(a, r)$ and $w \in N(a)$. Hence on integrating over $C(a, r)$, we have that $f_n^{(k)}(w) \to f^{(k)}(w)$ uniformly for $w \in N(a)$ as $n \to \infty$. But K is compact, and therefore covered by a finite number of the spheres $N(a)$, and so $f_n^{(k)} \to f^{(k)}$ uniformly on K.

Note 3.12. This theorem should be contrasted with the corresponding theorem for real functions on an interval. In the latter case the uniform convergence of the sequence of derivatives is needed as a hypothesis instead of forming part of the conclusion of the theorem.

EXERCISES FOR CHAPTER 3

1. Find the Taylor expansions of the following functions about the points indicated, and find the radius of convergence of the expansion in each case.

(i) $f(z) = (1 + z^2)^{-1}$ about $z = 0$ and $z = 2$.

(ii) $f(z) = \exp(z^3 - 1)$ about $z = 0$.

(iii) $f(z) = \tan z$ about $z = 0$, as far as terms in z^5.

2. Use Taylor's theorem, together with exercise 8 from Chapter 1 to show that if f is regular on $G \subset \mathbf{C}$, $a \in G$ and $|f|$ has a local maximum at a, then f must be constant.

In what circumstances can a non-constant regular function f have a local minimum of $|f|$?

3. The function f defined by

$$f(0) = 0, \qquad f(z) = z^2 \sin(1/z), \qquad z \neq 0$$

has zeros where $z = \pm 1/n\pi$, $n = 1, 2, 3, \ldots$.

Why does this not contradict theorem 3.4?

4. Find the Laurent expansions of the following functions in the regions indicated.

(i) $f(z) = z^{-1}(1 + z)^{-2}$ for $0 < |z| < 1$, for $|z| > 1$, and for $0 < |z + 1| < 1$.

(ii) $f(z) = z \sin(1/z)$ for $|z| > 0$.

(iii) $f(z) = \operatorname{cosec} z$ for $|z| > 0$, as far as terms in z^5.

5. Show that there is a Laurent expansion for the function

$$f(z) = (e^z - 1)^{-1},$$

of the form

$$\frac{1}{e^z - 1} = \frac{1}{z} + \sum_{n=0}^{\infty} a_n z^n,$$

valid for $0 < |z| < R$, and find the largest permissible value for R.

Find a_0, a_1, a_2 and a_3, and show that if $n \geqslant 1$, $a_{2n} = 0$.

6. Let $w \in \mathbf{C}$, and for $0 < |z| < \infty$, let $f(z) = \exp\{\frac{1}{2}w(z - 1/z)\}$. Expand f in a Laurent series about $z = 0$:

$$f(z) = \sum_{n=-\infty}^{\infty} J_n(w) z^n.$$

Show that

$$J_n(w) = (-1)^n J_{-n}(w) = \frac{1}{2\pi} \int_0^{2\pi} \cos(n\theta - w \sin\theta)\, d\theta.$$

[The functions J_n are called the Bessel functions of order n.]

7. Let f be regular on $S(0, R)$ for $R > 1$, and define g on $S(0, R) \setminus \{1\}$ by

$$g(z) = (1 - z)^{-1} f(z).$$

Let $\sum_{n=0}^{\infty} b_n z^n$ be the Taylor expansion of g about $z = 0$. Show that

$$b_n \to f(1) \quad \text{as} \quad n \to \infty.$$

8. Suppose that f is regular on $A(a; r, R)$, where $0 \leqslant r < R \leqslant \infty$, and let the Laurent expansion of f on A be

$$f(z) = \sum_{-\infty}^{\infty} a_n(z - a)^n.$$

Show that the series is absolutely convergent at all points of $A(a; r_1, R_1)$, where $0 \leqslant r_1 \leqslant r$, $R \leqslant R_1 \leqslant \infty$, and R_1 is given by

$$R_1 = (\limsup_{n \to +\infty} |a_n|^{1/n})^{-1}.$$

Find a corresponding formula for the inner radius r_1 in terms of the coefficients.

9. Let $G = \mathbf{C} \setminus \mathbf{Z}$, and define functions f_1 and f_2 on G by

$$f_1(z) = \frac{1}{z} + \sum_{n=1}^{\infty} \frac{2z}{z^2 - n^2} = \frac{1}{z} + \sum_{n=1}^{\infty} \left(\frac{1}{z - n} + \frac{1}{z + n}\right), \quad z \in G,$$

$$f_2(z) = \pi \cot \pi z, \quad z \in G.$$

Prove the following:

(a) Both functions are regular on G. (Use 3.11 for f_1.)

(b) Both functions have period 1; in other words:

$$f_i(z + 1) = f_i(z), \qquad i = 1, 2, \quad z \in G.$$

(c) $f_i(z) - (1/z) \to 0$ as $z \to 0$, $i = 1, 2$. (For f_2, use the Laurent series expansion about $z = 0$.)

(d) Both functions are uniformly bounded on any set of the form

$$\{z : z = x + iy, |y| \geqslant a > 0\}.$$

(For f_2, show that $f_2(x + iy) \to 1$ uniformly as $y \to \pm\infty$. For f_1, estimate the series by a sum of the form

$$\text{Const.} \sum_{n=1}^{\infty} \frac{|y|}{n^2 + y^2} \leqslant \text{Const.} \, |y| \int_0^{\infty} \frac{dt}{y^2 + t^2}.)$$

(e) $f_1(z) = f_2(z)$ for all $z \in G$. (Use (a)–(d) together with Liouville's theorem.) This result will be proved in a different way in Chapter 5.

10. Use the result of Q.9 to deduce the formula

$$\sin(\pi z) = \pi z \prod_{n=1}^{\infty} \left(1 - \frac{z^2}{n^2}\right),$$

where the product on the right-hand side is interpreted as the limit of the partial products $\prod_{n=1}^{N}$ as $N \to \infty$.

Hints: Let $0 < y < x < 1$. Show that the series for $\pi \cot \pi z$ can be integrated term-by-term from y to x. Then let $y \to 0$, and deduce the formula for real x. Finally justify the extension to complex z (theorem 3.4).

4

Zeros and Singularities of Regular Functions

In this chapter we use the power series expansions obtained in Chapter 3 to classify the behaviour of a function which is regular, or has an isolated singularity at a point. We also introduce the notion of the residue of a function at an isolated singularity, a concept which will be fundamental in Chapter 5.

§1. CLASSIFICATION OF ZEROS AND ISOLATED SINGULARITIES

Definition 4.1. We suppose that f is regular in an open set G, and that $a \in G$.

We say that a is a w-point of f if $f(a) = w$; in particular a is a zero of f if $f(a) = 0$. It follows from theorem 3.4 applied to $f_1 = f$, and $f_2 = w$ that if a is a w-point of f then either f is identically equal to w, or for some $r > 0$, $f(z) \neq w$ for $0 < |z - a| < r$. In other words each point a in the domain of regularity of a non-constant function is surrounded by an open sphere in which the value $f(a)$ is not repeated; we say simply that the w-points of a (non-constant) regular function are isolated.

Notice that the set of zeros may have a limit point in **C**, but that this cannot be a point of regularity—for instance $f(z) = \sin(\pi/z)$ is regular for $z \neq 0$ and has zeros at $z = \pm 1/k$, $k = 1, 2, \ldots$, but f is not regular at $z = 0$. If $S(a, R) \subset G$, and $f(a) = w$, then for all $z \in S(a, R)$,

$$f(z) = w + \sum_{n=1}^{\infty} a_n(z - a)^n.$$

If f is not identically equal to w, not all the coefficients a_n can vanish: the smallest integer $k \geqslant 1$ for which $a_k \neq 0$ is called the *order* of the w-point of f at a. We may then write

$$f(z) - w = (z - a)^k \sum_{n=k}^{\infty} a_n(z - a)^{n-k} = (z - a)^k g(z),$$

say, where g is regular at a and $g(a) \neq 0$. In particular if k is the order of a zero of f at a,

$$f(z) = (z - a)^k g(z) \quad \text{with} \quad g(a) \neq 0.$$

Since each

$$a_n = \frac{1}{n!} f^{(n)}(a),$$

the quickest way to determine the order is usually by considering derivatives, as in the following example.

Example 4.2. $f(z) = \cos z$ has a zero at $z = (2k + 1)(\pi/2)$ for each integer k. At each of these zeros, the derivative $f'(z) = -\sin z \neq 0$, so that the first non-vanishing coefficient is

$$a_1 = -\sin (2k + 1) \frac{\pi}{2} = (-1)^{k+1},$$

and the zeros are of the first order. On the other hand $f(0) = 1$, $f'(0) = 0$, $f''(0) = -1$, so that the order of the 1-point at $z = 0$ is two.

Definition 4.3. Let f be regular in an open set which contains the annulus $A(a; 0, R)$. If f is not regular at a, we say a is an *isolated singularity* of f. By theorem 3.7 we have the Laurent expansion $\sum_{-\infty}^{\infty} a_n(z - a)^n$ which converges to $f(z)$ on $A = A(a; 0, R)$.

We distinguish three possible cases.

(a) $a_n = 0$ for all $n < 0$. It follows that $f(z) = \sum_0^{\infty} a_n(z - a)^n$ on A, and if we define (or re-define, if f already has a value at a) the value of f to be a_0 at a, then the function thus obtained will be regular at a also. In this case we say f has a *removable singularity* at a. A necessary and sufficient condition for this to occur is given below as theorem 4.4.

(b) For some integer $k \geqslant 1$, $a_n = 0$ for all $n < -k$, while $a_{-k} \neq 0$. In this case $f(z) = \sum_{-k}^{\infty} a_n(z - a)^n$ on A, and

$$(z - a)^k f(z) = \sum_{n=0}^{\infty} a_{n-k}(z - a)^n = g(z),$$

say, is regular in $S(a, R)$. In addition $g(a) = a_{-k} \neq 0$. When this occurs, we say f has a *pole* of order k at a. By comparing definitions 4.1 and 4.3 it follows that f has a pole of order k at a if and only if the function $1/f$ has a zero of order k at a. Zeros and poles of order 1 are often referred to as simple zeros or poles respectively.

(c) $a_n \neq 0$ for infinitely many negative values of n. In this case we say f has an *essential singularity* at a.

Necessary and sufficient conditions for cases (b) and (c) to occur are given in theorem 4.5 and its corollary. Theorem 4.5 is known as the Casorati–Weierstrass theorem.

Theorem 4.4. *Let f be regular on $A = A(a; 0, R)$. Then f has a removable singularity at a if and only if $(z - a)f(z) \to 0$ as $z \to a$. In particular there is a removable singularity at a if f is bounded near a.*

Proof. If f has a removable singularity at a, $f(z)$ must approach a limit as $z \to a$, and hence $(z - a)f(z) \to 0$.

Conversely if $(z - a)f(z) \to 0$ as $z \to a$, so does $(z - a)^k f(z)$ for any integer $k \geqslant 1$. Suppose $n = -k \leqslant -1$. Then

$$a_n = \frac{1}{2\pi i} \int_{C(a,\rho)} \frac{f(z)\, dz}{(z - a)^{n+1}} \quad \text{for any} \quad 0 < \rho < R.$$

Given $\varepsilon > 0$, choose ρ so that $|(z - a)^k f(z)| < \varepsilon$ if $|z - a| \leqslant \rho$. Then

$$a_n = \frac{1}{2\pi i} \int_{C(a,\rho)} \frac{(z - a)^k f(z)\, dz}{(z - a)}, \quad \text{and} \quad |a_n| \leqslant \frac{1}{2\pi} \cdot \frac{2\pi\rho\varepsilon}{\rho} = \varepsilon,$$

Since ε is independent of n, we must have $a_n = 0$ for $n \leqslant -1$, and f has a removable singularity at a.

Theorem 4.5. *Let f be regular on $A(a; 0, R)$ and have an essential singularity at a. Then for any complex number w, and arbitrary $\varepsilon > 0$ and $\delta > 0$, there is a point z with $|z - a| < \delta$, and $|f(z) - w| < \varepsilon$. (This conclusion may be expressed by saying that on every $S(a, \delta)$, f takes values arbitrarily close to every complex number.)*

Proof. Suppose the conclusion of the theorem invalid. Then for some complex w_0 and $\alpha, \beta > 0$, $|f(z) - w_0| \geqslant \alpha$ on $S(a, \beta)$.

Consider $h(z) = (f(z) - w_0)^{-1}$. h is regular on $A(a; 0, \beta)$ and is bounded there by $1/\alpha$. From 4.4 it follows that h has a removable singularity at a; that is, we can suppose $h(a)$ defined so as to make h regular on $S(a, \beta)$. Then either $h(a) \neq 0$, or h has a zero of order k at a (h evidently cannot vanish identically on $S(a, \beta)$). It follows that

$$f(z) = w_0 + \frac{1}{h(z)}$$

is either regular at a, or has a pole of order k at a, contradicting the hypothesis that f has an essential singularity at a.

Corollary 4.6. *Let f be regular on $A(a; 0, R)$. Then f has a pole at a if and only if $|f(z)| \to \infty$ as $z \to a$.*

Proof. It was noted under (b) of definition 4.3 that if f has a pole of order k at a, then $f(z) = (z - a)^{-k}g(z)$ where g is regular and has a non-zero limit as $z \to a$. It follows that $|f(z)| \to \infty$ as $z \to a$.

Conversely if $|f(z)| \to \infty$ as $z \to a$, f cannot approach a finite limit, and hence there is not a removable singularity at a. For any complex w, and suitable $\delta > 0$,

$$|f(z)| \geqslant |w| + 1 \quad \text{for} \quad |z - a| < \delta,$$

and so $|f(z) - w| \geqslant 1$ if $z \in S(a, \delta)$. It follows that f cannot satisfy the conclusion of theorem 4.5, and so there is no essential singularity at a. Since f is evidently not regular at a, the only remaining possibility is that f has a pole at a.

We now examine some particular examples of functions in order to classify their (isolated) singularities.

Example 4.7. (i) $f(z) = (1 + z^2)^{-1}$ has poles of order 1 (simple poles) at $z = \pm i$. $f(z) = (1 + z)^{-2}$ has a pole of order 2 at $z = -1$.

(ii) $f(z) = \cot z$ has a simple pole at $z = k\pi$, and a simple zero at $(k + \frac{1}{2})\pi$ for each integer k. $g(z) = z \cot z$ has the same zeros and poles as f, except at $z = 0$; however,

$$g(z) = \frac{z}{\sin z} \cos z,$$

which $\to 1$ as $z \to 0$, so g has a removable singularity at 0. If $g(0)$ is defined as 1, g will be regular at $z = 0$.

(iii)

$$f(z) = \exp\left(\frac{1}{z}\right) = \sum_{k=0}^{\infty} \frac{1}{k!} z^{-k}$$

has an essential singularity at $z = 0$. For every complex $w \neq 0$, and every $\delta > 0$, it follows from the properties listed in 1.8 that there is a value of z, $|z| < \delta$, with $f(z) = w$. The conclusion of theorem 4.5 is then valid in the stronger sense that, with at most one exceptional value of w, $f(z)$ may be made equal to w, instead of merely within ε of w. This is a particular instance of a famous (and difficult) theorem due to Picard which asserts that this behaviour occurs at every essential singularity of a regular function.

It now becomes convenient to augment the complex plane by the addition of a single 'point at infinity'. Denote by ∞ any element which is not already a complex number, and denote by $\mathbf{C}^*$ the set $\mathbf{C} \cup \{\infty\}$. The points of $\mathbf{C}^*$ may be made to correspond in a very natural way with the points on the surface $\mathbf{S}^2$ of the unit sphere in $\mathbf{R}^3$ (three-dimensional real Euclidean space). The point

$$(r \cos\theta + ir \sin\theta) \in \mathbf{C} \quad \text{corresponds to} \quad \left(\frac{2r\cos\theta}{r^2+1}, \frac{2r\sin\theta}{r^2+1}, \frac{r^2-1}{r^2+1}\right),$$

while ∞ corresponds to $(0, 0, 1)$. Geometrically, the correspondence is given by stereographic projection from $\mathbf{S}^2$ onto $\mathbf{C}$, using the point $(0, 0, 1)$ (the 'North Pole' of $\mathbf{S}^2$) as the centre for the projection. We define the open sets in $\mathbf{C}^*$ in such a way as to make this correspondence and its inverse continuous. More explicitly we have

Definition 4.8. A set $G \subset \mathbf{C}^*$ is open if either

(a) $\infty \notin G$, and G is open in $\mathbf{C}$, or

(b) $\infty \in G$, $G \setminus \{\infty\}$ is open in $\mathbf{C}$, and for some $R > 0$,

$$\{z: |z| > R\} \subset G.$$

Definition 4.9. Suppose f is a function regular on an open set $G \subset \mathbf{C}$, where $G \cup \{\infty\}$ is open in $\mathbf{C}^*$. If $z^{-1} \in G$, define

$$\tilde{f}(z) = f(z^{-1}):$$

it follows from (b) of definition 4.8 that $\tilde{f}$ is regular on some $S(0, r) \setminus \{0\}$.

We classify the behaviour of f at ∞ by making it correspond to that of $\tilde{f}$ at 0.

More explicitly we say that f has a removable singularity at ∞ if $\tilde{f}$ has a removable singularity at 0. In particular, if $\tilde{f}(0)$ is defined so that $\tilde{f}$ is regular at 0, then we define $f(\infty) = \tilde{f}(0)$, and say that f is regular at ∞. It should be observed that regularity at ∞ is not defined in terms of any concept of differentiability at ∞.

Continuing in this way, we say that f has a zero of order k, or a pole of order k, or an essential singularity at ∞ if the corresponding statement is true of $\tilde{f}$ at 0.

Examples 4.10. (i) A polynomial of degree k has a pole of order k at ∞. For if

$$f(z) = a_0 + a_1 z + \cdots + a_k z^k, \quad a_k \neq 0 \quad \text{for} \quad z \in \mathbf{C},$$

we have $\tilde{f}(z) = f(z^{-1}) = a_k z^{-k} + \cdots + a_0$, and $\tilde{f}$ has a pole of order k at 0.

(ii) Any function which is regular on $\mathbf{C}$ and not a polynomial has an essential singularity at ∞.

For it follows from Taylor's theorem that $f(z) = \sum_{k=0}^{\infty} a_k z^k$, where the series is convergent for all $z \in \mathbf{C}$, and infinitely many a_k are $\neq 0$, since f is not a polynomial. Then $\tilde{f}(z) = \sum_{k=0}^{\infty} a_k z^{-k}$, which has an essential singularity at 0.

(iii) $f(z) = z(1 + z^2)^{-1}$ has a simple zero (a zero of order 1) at ∞, since $\tilde{f}(z)$ also equals $z(1 + z^2)^{-1}$, which has a simple zero at $z = 0$.

(iv) $f(z) = z - z^{-1}$ has simple zeros at $z = \pm 1$, and simple poles at $z = 0$ and ∞.

Theorem 4.11. (a) *Suppose f is regular on* $\mathbf{C}^*$. *Then f is constant.*

(b) *Suppose f is regular on* $\mathbf{C}^*$ *except for a finite number of poles at points of* $\mathbf{C}^*$. *Then f is a rational function; that is for some polynomials P, Q, $f(z) = P(z)/Q(z)$.*

Proof. (a) If f is regular at ∞, $f(z)$ must tend to a finite limit as $z \to \infty$, and in particular $f(z)$ must be bounded for $|z| > R$, say. But f is continuous on $\bar{S}(0, R)$ and is therefore bounded on the whole of **C**. Liouvilles' theorem (2.11) now shows that f is constant.

(b) Suppose f has poles at $a_1, a_2, \ldots, a_n \in \mathbf{C}$, of orders

$$k_1, k_2, \ldots, k_n.$$

Then $g(z) = f(z)(z - a_1)^{k_1}(z - a_2)^{k_2} \ldots (z - a_n)^{k_n}$ is regular on **C**, and has at worst a pole at ∞. It follows from (ii) of 4.10 that g is a polynomial, and hence that f is rational.

§2. RESIDUES

We now return to the consideration of isolated singularities at points of **C**, to define the notion of a residue which will be used extensively in Chapter 5.

Definition 4.12. Let $a \in \mathbf{C}$, and f be regular on $A(a; 0, R)$ for some $R > 0$. By Laurent's theorem, $f(z) = \sum_{n=-\infty}^{\infty} a_n(z - a)^n$, where the series converges absolutely at all points of A. The coefficient a_{-1} is called the residue of f at a, and will be denoted by $R(f, a)$.

This very important definition is motivated by the following considerations. The coefficient a_n is given by

$$\frac{1}{2\pi i} \int_{C(a,\rho)} \frac{f(z)\, dz}{(z - a)^{n+1}},$$

and in particular if $n = -1$,

$$a_{-1} = \frac{1}{2\pi i} \int_{C(a,\rho)} f(z)\, dz,$$

for any ρ, $0 < \rho < R$. This may be written $\int_{C(a,\rho)} f(z)\, dz = 2\pi i a_{-1}$; in other words the integral of f around a circle centred on an isolated singularity may be found by evaluating the residue, and multiplying by $2\pi i$. The residue theorem of Chapter 5 is an extension of this fact to a wider class of curves and functions with more than one singularity. Some methods of evaluating residues for particular classes of functions are given below.

Examples 4.13. (a) Let f have a simple pole at $z = a$. Then

$$f(z) = \sum_{-1}^{\infty} a_n(z - a)^n$$

in some $A(a; 0, R)$, and $(z - a)f(z) = \sum_0^\infty a_{n-1}(z - a)^n$.

Hence

$$R(f, a) = a_{-1} = \lim_{z \to a} \{(z - a)f(z)\}.$$

For instance if

$$f(z) = \frac{1}{(z - 1)(z + 2)},$$

f has simple poles at 1 and -2, with residues $\frac{1}{3}$ and $-\frac{1}{3}$ respectively.

This situation occurs in particular when f is given in the form $f(z) = g(z)/h(z)$, where both g and h are regular at a, and h has a simple zero at a, while $g(a) \neq 0$.

In this case

$$R(f, a) = \lim_{z \to a} \{(z - a)f(z)\} = g(a) \lim_{z \to a} \left\{\frac{z - a}{h(z) - h(a)}\right\} = \frac{g(a)}{h'(a)},$$

where $h'(a) \neq 0$ since the zero of h at a is simple.

This is often of help in more complicated cases: for instance if $f(z) = z^2/(e^z - 1)$, then f has a simple pole at $z = 2k\pi i$ for integers $k \neq 0$ (but a simple zero at $z = 0$). Then

$$R(f, 2\pi ki) = \frac{(2k\pi i)^2}{e^{2k\pi i}} = -4\pi^2k^2, \qquad k \neq 0.$$

(b) The function $f(z) = \pi \cot \pi z$ will be needed for the investigation of series in Chapter 5. It has a simple pole at $z = k$ for any integer k, and a simple zero at $z = k + \frac{1}{2}$.

Since

$$f(z) = \frac{\pi \cos \pi z}{\sin \pi z}, \quad \text{and} \quad (\sin \pi z)' = \pi \cos \pi z,$$

the residue of f at each pole is 1.

(c) In the case of poles of higher order, or of essential singularities, it is usually advisable to find the Laurent coefficients from first principles.

For example if $f(z) = e^z(z-1)^{-2}$, then f has a double pole at $z = 1$. In this case

$$f(z) = (z-1)^{-2}e \cdot e^{z-1} = e\left(\frac{1}{(z-1)^2} + \frac{1}{(z-1)} + \frac{1}{2!} + \cdots\right),$$

and $R(f, 1) = e$.

(d) The function $f(z) = \cos(1/z)$ has an essential singularity at $z = 0$. From the expansion

$$f(z) = \sum_{k=0}^{\infty} \frac{(-1)^k}{(2k)!} z^{-2k}$$

it follows that $R(f, 0) = 0$.

Similarly $f(z) = z\cos(1/z)$ has an essential singularity at $z = 0$, and $R(f, 0) = -\frac{1}{2}$.

The function $f(z) = \exp\{z + 1/z\}$ also has an essential singularity at $z = 0$. It follows on multiplying together the series expansions for e^z and $e^{1/z}$, that

$$R(f, 0) = \sum_{k=0}^{\infty} \frac{1}{k!(k+1)!}$$

EXERCISES FOR CHAPTER 4

1. For each of the functions specified by the following formulae, find the location and orders of all zeros, including those at ∞.

(i) $f(z) = z^2 + 1$,
(ii) $f(z) = z^6 - 2z^3 + 1$,
(iii) $f(z) = \left(\dfrac{\sin z}{z}\right)^2$,
(iv) $f(z) = \dfrac{z}{z^3 + 1}$,
(v) $f(z) = 1 - e^{-1/z}$.

2. For each of the functions specified by the following formulae, find and classify all isolated singularities, including those at ∞.

(i) $f(z) = z^2 + z + 1$,
(ii) $f(z) = \dfrac{1}{z}(z - \frac{1}{4}i)(1 + \frac{1}{4}iz)$,
(iii) $f(z) = \sin z + \cos z$,
(iv) $f(z) = \operatorname{cosec}^2 z$,
(v) $f(z) = z^2 e^{-1/z^2}$.

3. Show that if f has a pole of order k at a, then

$$R(f, a) = \frac{1}{(k-1)!}\left[\frac{d^{k-1}}{dz^{k-1}}\{(z-a)^k f(z)\}\right]_{z=a}.$$

4. Find the residues of the functions given by the following formulae at each of their finite isolated singularities.

(i) $f(z) = \dfrac{z^2 + 1}{z^2 - 1}$,

(ii) $f(z) = \dfrac{z^2 + 1}{(z - 1)^2}$,

(iii) $f(z) = \dfrac{\cos z}{z^2}$,

(iv) $f(z) = \operatorname{cosec} z$,

(v) $f(z) = \sin(1/z)$,

(vi) $f(z) = z^2 e^{-1/z^2}$.

5. Show that if $a_1, a_2, \ldots, a_n$ are distinct complex numbers, and P is a polynomial whose degree is less than or equal to $n - 1$, then

$$f(z) = \frac{P(z)}{(z - a_1)(z - a_2)\ldots(z - a_n)} = \sum_{j=1}^{n} \frac{R(f, a_j)}{z - a_j}.$$

[This is the simplest case of the 'partial fractions' decomposition of f, familiar from elementary algebra.]

6. Show that if f has an essential singularity at a point a of $\mathbf{C}$, then so does f^2. If f and g have essential singularities at a, need fg have one?

Let f be regular on $A = S(a, R) \setminus \{a\}$, and have an isolated singularity at a. What conditions are needed on a regular function g to ensure that fg has the same behaviour as f at a? Can you say anything about the corresponding problem for $g \circ f$, when g is a polynomial?

7. Let a be a point of an open set G, and suppose f is regular on $G \setminus \{a\}$, and has a simple pole at a. Suppose $\bar{S}(a, R) \subset G$.

Show that

$$\frac{1}{2\pi i} \int_{C(a,R)} f(z)\, dz = R(f, a).$$

(Use the fact that $f(z) - R(f, a)(z - a)^{-1}$ is regular on G, together with lemma 1.22. This simple formula contains the Cauchy integral formula 2.5 as a special case, and will in turn be extensively generalized in the residue theorem of Chapter 5.)

5

The Residue Theorem

This chapter forms the culmination of the elementary part of complex analysis. In it we apply the full strength of Cauchy's theorem to the integration of regular functions around general closed paths, instead of the circular paths used throughout chapters 3 and 4. We begin with a method of distinguishing between the various components into which a closed path divides the plane by assigning an integer called the topological index (or winding number) to each component.

§1. THE TOPOLOGICAL INDEX

Definition 5.1. Let γ be a curve; that is γ maps $[0, 1]$ continuously into $\mathbf{C}$.

The track γ^* is a compact subset of $\mathbf{C}$ as was noted in 1.15 (ii), and in particular the complement $\mathbf{C} \setminus \gamma^*$ is open and so may be expressed as the union of a number of connected components as in 1.18.

In addition, since γ^* is bounded, for some $R > 0$ the set

$$E_R = \{z : |z| > R\}$$

is disjoint from γ^*. It now follows from the fact that E_R is connected, that E_R is contained in exactly one component of $\mathbf{C} \setminus \gamma^*$. This component is unbounded and will be called the exterior of γ, denoted $E(\gamma)$.

Any other component of $\mathbf{C} \setminus \gamma^*$ is contained in $\bar{S}(0, R)$ and is therefore bounded. The examples below show that $\mathbf{C} \setminus \gamma^*$ may have no bounded components, or a finite number, or a denumerably infinite number.

Examples 5.2. (i) Let $\gamma(t) = t$ for $t \in [0, 1]$. Then $\gamma^* = [0, 1] \subset \mathbf{C}$, and the exterior $E(\gamma) = \mathbf{C} \setminus [0, 1]$. There are no bounded components.

(ii) Let $\gamma(t) = a + r\, e^{2\pi i t}$, $t \in [0, 1]$. Here

$$\gamma^* = C(a, r), \qquad E(\gamma) = \{z\colon |z - a| > r\},$$

and the only bounded component is $S(a, r)$.

(iii) Let $\gamma(t) = t + it \sin[\pi/(2t)]$ for $0 < t \leqslant \frac{1}{2}$, $\gamma(0) = 0$, $\gamma(t) = 1 - t$ for $\frac{1}{2} < t \leqslant 1$. For this closed curve, $\mathbf{C} \setminus \gamma^*$ has infinitely many components.

In order to be able to distinguish between the components of $\mathbf{C} \setminus \gamma^*$ in the case where γ is a closed path, we now introduce an integer valued function of the point $w \notin \gamma^*$.

Theorem 5.3. *Let γ be a closed path, and w a point not on γ^*. Then $\int_\gamma (dz)/(z - w)$ is an integer multiple of $2\pi i$.*

Proof (Ahlfors). Since $w \notin \gamma^*$,

$$\theta(t) = \int_0^t \frac{\gamma'(u)\, du}{\gamma(u) - w}$$

is well defined, and $\theta(0) = 0$, while $\theta(1) = \int_\gamma (dz)/(z - w)$.

Since γ is assumed to be a closed path, $\gamma(0) = \gamma(1)$, and γ' is continuous on $[0, 1]$, except for a finite set of points $0 = t_1 < t_2 < \cdots < t_n = 1$ at which left- and right-hand derivatives exist, but may be unequal.

Then for all $t \neq t_k$, $k = 1, 2, \ldots, n$,

$$\theta'(t) = \frac{\gamma'(t)}{\gamma(t) - w}.$$

We rewrite this as

$$\gamma'(t) - \gamma(t)\theta'(t) + w\theta'(t) = 0, \qquad t \neq t_k,$$

and hence

$$\frac{d}{dt}\{(\gamma(t) - w)\, e^{-\theta(t)}\} = \gamma'(t)\, e^{-\theta(t)} - \theta'(t)\, e^{-\theta(t)}(\gamma(t) - w) = 0.$$

Hence $(\gamma(t) - w)\, e^{-\theta(t)}$ is constant on each interval (t_{k-1}, t_k): however, both γ and θ are continuous functions of t, and hence $(\gamma(t) - w)\, e^{-\theta(t)}$ is constant on $[0, 1]$.

In particular

$$(\gamma(0) - w)\, e^{-\theta(0)} = (\gamma(1) - w)\, e^{-\theta(1)},$$

and since $\gamma(1) = \gamma(0) \neq w$, we have $e^{-\theta(1)} = e^{-\theta(0)} = 1$.

Hence $\theta(1)$ is an integer multiple of $2\pi i$ as required.

Definition 5.4. If γ is a closed path and $w \notin \gamma^*$, then the integer $[1/(2\pi i)] \int_\gamma (dz)/(z - w)$ is called the topological index of γ with respect to w. (Alternative names are simply 'index' or 'winding number'.) We denote it by $n(\gamma, w)$. Notice that lemma 1.22 now says that if $\gamma(t) = a + r\, e^{2\pi i t}$, so that $\gamma^* = C(a, r)$, then

$$n(\gamma, w) = \begin{cases} 1 & \text{if } |w - a| < r \\ 0 & \text{if } |w - a| > r. \end{cases}$$

The notion of topological index may be extended to all closed curves by a process of segmental approximation, but we shall have no need of this. The index $n(\gamma, w)$ has the properties listed in the following two results (5.5 and 5.7).

Lemma 5.5. *Let γ be a closed path and $w \notin \gamma^*$. Then*

(i) $n(-\gamma, w) = -n(\gamma, w)$.

(ii) *As a function of w, $n(\gamma, w)$ is constant in each component of $\mathbf{C} \setminus \gamma^*$.*

(iii) *If $w \in E(\gamma)$, $n(\gamma, w) = 0$.*

(iv) *If $\gamma^* \subset G$, a starred open set in $\mathbf{C}$, and $w \notin G$, then $n(\gamma, w) = 0$.*

Proof. (i) This is simply a special case of (ii) of 1.20.

(ii) Since γ^* is closed, we choose $r > 0$ so that $S(w, r) \cap \gamma^* = \varnothing$. If now we take a point $w' \in S(w, \tfrac{1}{2}r)$, and let z be any point of γ^*, then

$$|z - w| \geqslant r \quad \text{and} \quad |z - w'| \geqslant \tfrac{1}{2}r.$$

It follows that

$$|n(\gamma, w) - n(\gamma, w')| = \left|\frac{1}{2\pi i} \int_\gamma \left(\frac{1}{z - w} - \frac{1}{z - w'}\right) dz\right|$$
$$\leqslant \frac{|w - w'| L(\gamma)}{2\pi r(\tfrac{1}{2}r)},$$

which tends to zero as w' approaches w. Hence $n(\gamma, w)$ is a continuous function of $w \in \mathbf{C} \setminus \gamma^*$, and since it is integer valued, it now follows from 1.17 that it must be constant on each component of $\mathbf{C} \setminus \gamma^*$.

(iii) This is proved in similar manner to (ii). Suppose $\gamma^* \subset S(0, R)$, and that $|w| > R$, whence $w \in E(\gamma)$ as in 5.1. Then

$$|n(\gamma, w)| \leqslant \frac{1}{2\pi} \frac{L(\gamma)}{|w| - R}$$

and so $|n(\gamma, w)| < 1$ for $|w|$ sufficiently large. But $n(\gamma, w)$ must be an integer, and hence must be zero for $|w|$ sufficiently large. Hence by (ii), $n(\gamma, w) = 0$ for all $w \in E(\gamma)$.

(iv) Since $w \notin G$, $f(z) = 1/(z - w)$ is regular in G; moreover G is starred by hypothesis. Hence

$$\int_\gamma \frac{dz}{z - w} = 0$$

by Cauchy's theorem (2.3).

We need now a tool to enable us to evaluate the index in cases when it is not zero. We do this by giving a meaning to the direction in which a path crosses a line segment, and using this to evaluate the index in one component in terms of that in a neighbouring one.

Definition 5.6. Let γ be a closed curve, and a, b be distinct points of $\mathbf{C} \setminus \gamma^*$.

Then we shall say that γ crosses the directed segment $[a, b]$ in the positive direction if there is a point $t_0 \in [0, 1)$ such that

(i) $\gamma(t) \in [a, b]$ if and only if $t = t_0$, and

(ii) there is a $\delta > 0$ for which

$$\operatorname{Im}\left\{\frac{\gamma(t) - a}{b - a}\right\} > 0 \quad \text{if} \quad t \in (t_0, t_0 + \delta),$$

and

$$\operatorname{Im}\left\{\frac{\gamma(t) + a}{b - a}\right\} < 0 \quad \text{if} \quad t \in (t_0 - \delta, t_0)$$

(or $(1 - \delta, 1)$ if $t_0 = 0$).

Intuitively speaking, if we rotate the plane so that a and b are on the real axis with $a < b$, then the curve crosses $[a, b]$ in the positive direction if it crosses from the lower half plane $\{z: \operatorname{Im} z < 0\}$ to the upper half plane $\{z: \operatorname{Im} z > 0\}$ as t increases through t_0. If the opposite curve $-\gamma$ crosses $[a, b]$ in the positive direction, we say γ crosses $[a, b]$ in the negative direction.

Theorem 5.7. *Let γ be a closed path, let a, b be distinct points of $\mathbf{C} \setminus \gamma^*$, and suppose that γ crosses $[a, b]$ in the positive direction.*
Then $n(\gamma, a) = n(\gamma, b) + 1$.

Proof. We suppose that $t_0 \in [0, 1)$ and $\delta > 0$ are as in definition 5.6. Let $c = \gamma(t_0)$. Choose $r > 0$ with

$|\gamma(t_0 + \delta) - c|$, $|\gamma(t_0 - \delta) - c|$, $|b - c|$ and $|a - c|$ all $>r$.
Let

$$t_1 = \inf\{t: t > t_0, \gamma(t) \in C(c, r)\},$$
$$t_2 = \sup\{t: t < t_0, \gamma(t) \in C(c, r)\},$$

and let $z_1 = \gamma(t_1)$, $z_2 = \gamma(t_2)$ (see figure 5.1). Since γ is continuous, both z_1 and z_2 are on $C(c, r)$ but not on $[a, b]$.

Denote by I that subarc of γ from z_2 to z_1 which contains c, and by J the remainder of γ from z_1 back to z_2. Denote by C_1 the subarc of $C = C(c, r)$ from z_2 to z_1 which intersects $[c, b]$, and by C_2 the remain-

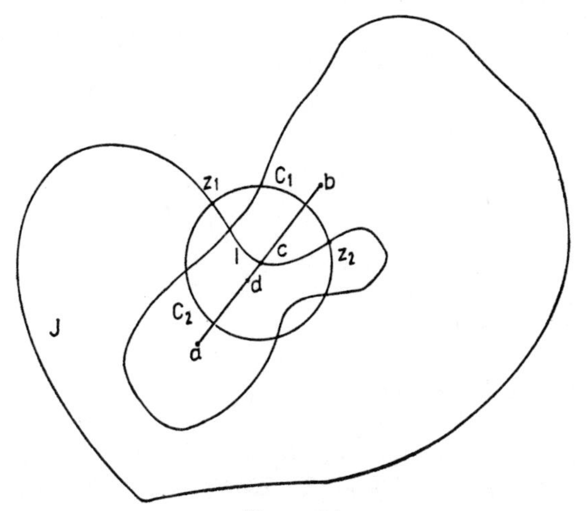

Figure 5.1

der of C from z_2 back to z_1. Let d be a point of $[a, c]$, with $|d - c| < r$. We are now in a position to compute various values of n.

Firstly

$$n(\gamma, a) = \frac{1}{2\pi i}\int_\gamma \frac{dz}{z - a} = \frac{1}{2\pi i}\int_I \frac{dz}{z - a} + \frac{1}{2\pi i}\int_J \frac{dz}{z - a},$$

and if γ_1 denotes $C_1 \cup (-I)$, γ_2 denotes $I \cup C_2$,

$$n(\gamma_1, a) = \frac{1}{2\pi i}\int_{C_1} \frac{dz}{z - a} - \frac{1}{2\pi i}\int_I \frac{dz}{z - a},$$

But $\gamma_1^* \subset \bar{S}(c, r)$, while $a \notin \bar{S}(c, r)$ whence $a \in E(\gamma_1)$ and $n(\gamma_1, a) = 0$. Hence

$$\begin{aligned} n(\gamma, a) &= n(\gamma, a) + n(\gamma_1, a) \\ &= \frac{1}{2\pi i}\int_J \frac{dz}{z - a} + \frac{1}{2\pi i}\int_{C_1} \frac{dz}{z - a} = n(J \cup C_1, a). \end{aligned}$$

But the segment $[a, d]$ is disjoint from $J \cup C_1$, and hence

$$n(J \cup C_1, a) = n(J \cup C_1, d).$$

In an exactly similar way we may show that

$$n(\gamma, b) = n(\gamma, b) - n(\gamma_2, b) = n(J \cup -C_2, b) = n(J \cup -C_2, d).$$

It follows that

$$n(\gamma, a) - n(\gamma, b) = n(C_1 \cup C_2, d).$$

The point $\qquad C_1 \cap [c, b]$ is $c + r\dfrac{b - a}{|b - a|}$,

so that the next point in anticlockwise order on the circle $C(c, r)$ (see 1.16 (iv)) is

$$c + ir\frac{b - a}{|b - a|}.$$

This point satisfies

$$\operatorname{Im}\left(\frac{z - a}{b - a}\right) > 0,$$

so that $C_1 \cup C_2$ is simply $C(c, r)$ described in the anticlockwise direction. Since $d \in S(c, r)$, lemma 1.22 shows that $n(C_1 \cup C_2, d) = 1$, which completes the proof.

We now illustrate the use of this lemma by means of some examples.

Examples 5.8. (i) Let γ be the square with sides

$$S_j = [i^{j-1}R(1 + i), i^jR(1 + i)],$$

for some $R > 0, j = 1, 2, 3, 4$.

Then the only bounded component of $\mathbf{C} \setminus \gamma^*$ is

$$\{z: z = x + iy, |x| < R, |y| < R\},$$

and since the curve crosses the segment $[0, 2R]$ in the positive direction, it follows that

$$n(\gamma, 0) = n(\gamma, 2R) + 1 = 1.$$

(ii) Let γ be the perimeter of a semicircle,

$$\gamma = [-R, R] \cup \gamma_R, \quad \text{where} \quad \gamma_R(t) = R\, e^{\pi it}, \quad 0 \leqslant t \leqslant 1.$$

Again $\mathbf{C} \setminus \gamma^*$ has only one bounded component, and taking $\frac{1}{2}iR$ as a representative point we see that

$$n(\gamma, \tfrac{1}{2}iR) = n(\gamma, -\tfrac{1}{2}iR) + 1 = 0 + 1 = 1.$$

(iii) Let

$$\gamma(t) = (1 + 2\cos 2\pi t)\, e^{2\pi it}, \qquad 0 \leqslant t \leqslant 1.$$

$\gamma(t)$ is real when $t = 0$ or $\frac{1}{2}$ giving $\gamma(t) = 3, 1$ respectively, and also when $t = \frac{1}{3}, \frac{2}{3}$ giving $\gamma(t) = 0$, and at $t = 0$ and $\frac{1}{2}$ the curve crosses the real axis in the positive direction. Also we have $n(\gamma, 4) = 0$ (since $\operatorname{Re} \gamma(t) \leqslant 3$ for all t) and so

$$n(\gamma, 2) = n(\gamma, 4) + 1 = 1, \quad \text{and} \quad n(\gamma, \tfrac{1}{2}) = n(\gamma, 2) + 1 = 2.$$

§2. THE RESIDUE THEOREM

We now use our knowledge of residues and topological indices to prove and apply the following theorem which includes many of the theorems of Chapters 2 and 3 as special cases.

Theorem 5.9 (The residue theorem). *Let G be a starred open set in* $\mathbf{C}$, *and A a finite subset of G.*

Let f be regular on $G \setminus A$, and let γ be a closed path with $\gamma^* \subset G \setminus A$.

Then

$$\int_\gamma f(z)\,dz = 2\pi i \sum_{a\in A} n(\gamma, a)R(f, a).$$

Proof. Let $\{a_1, a_2, \ldots, a_n\}$ be the points of A.

By the note (i) in 3.9, we know that if f_k denotes the principal part of f at a_k ($k = 1, 2, \ldots, n$), then f_k is regular on $\mathbf{C} \setminus \{a_k\}$ and $f - f_k$ is regular at a_k. Also f_k has a series expansion

$$f_k(z) = \sum_{j=-\infty}^{-1} a_{k,j}(z - a_k)^j,$$

where $a_{k,-1}$ is the residue of f at a_k. Let

$$g(z) = f(z) - (f_1(z) + f_2(z) + \cdots + f_n(z)).$$

By our construction, g is regular at each a_k, $k = 1, 2, \ldots, n$, and hence on all of G. It follows from Cauchy's theorem (2.3), that

$$\int_\gamma g(z)\,dz = 0,$$

or equivalently,

$$\int_\gamma f(z)\,dz = \sum_{k=1}^{n} \int_\gamma f_k(z)\,dz.$$

Since each $a_k \notin \gamma^*$, there is a $\delta_k > 0$ with $\gamma^* \cap S(a_k, \delta_k) = \varnothing$. But the series representation of f_k converges uniformly off $S(a_k, \delta_k)$, and so in particular on γ^*.

It follows that to evaluate $\int_\gamma f_k(z)\,dz$, we may integrate term by term and obtain

$$\sum_{j=-\infty}^{-1} a_{k,j} \int_\gamma (z - a_k)^j\,dz.$$

If $j \neq -1$,

$$\int_\gamma (z - a_k)^j\,dz = \frac{1}{j+1}(z - a_k)^{j+1}\Big|_\gamma = 0,$$

while if $j = -1$,

$$\int_\gamma (z - a_k)^{-1}\,dz = 2\pi i n(\gamma, a_k).$$

It follows that

$$\int_\gamma f(z)\,dz = \sum_{k=1}^{n} a_{k,-1}(2\pi i n(\gamma, a_k)) = 2\pi i \sum_{k=1}^{n} n(\gamma, a_k)R(f, a_k).$$

We now illustrate the practical uses of the residue theorem by means of a number of examples.

Example 5.10. The substitution $z = e^{i\theta}$, $0 \leqslant \theta \leqslant 2\pi$ may be used to convert an integral involving trigonometric functions on the interval $[0, 2\pi]$, to one involving a rational function around the unit circle.

For instance, suppose we have to evaluate

$$\int_0^{2\pi} \frac{\cos n\theta}{a + \cos\theta}\,d\theta,$$

where n is a positive integer, and a is a real number, which must be taken >1 to ensure that the denominator does not vanish.

Let I_n be the required integral, and

$$J_n = \int_0^{2\pi} \frac{\sin n\theta}{a + \cos\theta}\,d\theta.$$

(Since J_n is the integral of an odd periodic function, it must equal zero for all n, but we do not need this fact at once.)

Then

$$\begin{aligned} I_n + iJ_n &= \int_0^{2\pi} \frac{e^{in\theta}\,d\theta}{a + \cos\theta} \\ &= \int_{C(0,1)} \frac{z^n}{a + \frac{1}{2}(z + 1/z)}\frac{dz}{iz} = \frac{2}{i}\int_{C(0,1)} \frac{z^n\,dz}{z^2 + 2az + 1}. \end{aligned}$$

Let

$$f(z) = \frac{z^n}{z^2 + 2az + 1};$$

then f has simple poles at the points

$$z = a_1 = -a + \sqrt{(a^2 - 1)} \quad \text{and at} \quad z = a_2 = -a - \sqrt{(a^2 - 1)}.$$

Further $a_2 < -1$, and since $a_1 a_2 = 1$ we have $-1 < a_1 < 0$. In particular by 1.22, the index of a_2 with respect to $C(0, 1)$ is 0, and that of a_1 is 1.

The residue of f at a_1 is

$$R(f, a_1) = \frac{(a_1)^n}{2a_1 + 2a} = \frac{(a_1)^n}{2\sqrt{(a^2 - 1)}}.$$

Applying the residue theorem to f, with $G = \mathbf{C}$, $A = \{a_1, a_2\}$, $\gamma = C(0, 1)$ now gives

$$\int_{C(0,1)} f(z)\, dz = 2\pi i \sum_{k=1}^{2} R(f, a_k) n(\gamma, a_k) = 2\pi i \cdot \frac{(a_1)^n}{2\sqrt{(a^2 - 1)}}$$

Hence

$$I_n + iJ_n = \frac{2}{i} \cdot \frac{2\pi i (a_1)^n}{2\sqrt{(a^2 - 1)}}, \quad \text{or}$$

$$I_n = \frac{2\pi[-a + \sqrt{(a^2 - 1)}]^n}{\sqrt{(a^2 - 1)}}, \qquad J_n = 0.$$

Example 5.11. Suppose $f(z)$ is a rational function of z with no poles on the real axis, and that we wish to evaluate $\int_{-\infty}^{\infty} f(x)\, dx$. If $f(z) = P(z)/Q(z)$, where P, Q are polynomials with no common factors, we must assume that the degree of Q exceeds that of P by at least two in order to ensure the convergence of the integral: we are also assuming that $Q(x) \neq 0$ for real x.

We consider $\int_{-\infty}^{\infty} f(x)\, dx$ as the limit as $R \to \infty$ of $\int_{-R}^{R} f(x)\, dx$, and in order to have a closed curve at our disposal we consider the semicircular path $\gamma = [-R, R] \cup \gamma_R$, as in (ii) of 5.8. Now Q will have finite number of zeros in the upper half-plane $\{z: \operatorname{Im} z > 0\}$, say at $z = \{a_1, a_2, \ldots, a_n\}$, and these will be poles of f.

If we now take

$$R > \max(|a_1|, |a_2|, \ldots, |a_n|),$$

then as in (ii) of 5.8, $n(\gamma, a_k) = 1$ for $k = 1, 2, \ldots, n$ while $n(\gamma, a) = 0$ for points a in the lower half plane.

It follows from the residue theorem, applied to $G = \mathbf{C}$, $A = \{\text{poles of } f\}$, γ as above, that

$$\int_{-R}^{R} f(x)\, dx + \int_{\gamma_R} f(z)\, dz = 2\pi i \sum_{k=1}^{n} R(f, a_k).$$

The condition on the degrees of P and Q ensures the existence of a constant M for which $|f(z)| \leqslant M|z|^{-2}$ for $|z|$ sufficiently large, and hence that

$$\left|\int_{\gamma_R} f(z)\, dz\right| \leqslant \pi R \cdot MR^{-2} \to 0 \quad \text{as} \quad R \to \infty.$$

Hence

$$\int_{-\infty}^{\infty} f(x)\, dx = 2\pi i \sum_{k=1}^{n} R(f, a_k),$$

when the a_k are the poles of f, with $\text{Im}\,(a_k) > 0$.

Suppose in particular that we have to evaluate $\int_{-\infty}^{\infty} dx/(1 + x^2)^2$. The above conditions on $f(z) = (1 + z^2)^{-2}$ are obviously satisfied, and f has double poles at $z = \pm i$. The residue of

$$f(z) = \frac{1}{(z - i)^2(2i + (z - i))^2} = -\frac{1}{4(z - i)^2}\left(1 + \frac{1}{2i}(z - i)\right)^{-2}$$

$$= -\frac{1}{4(z - i)^2} + \frac{1}{4i(z - i)} + \cdots,$$

at $z = i$ is $1/(4i)$.

Then by the above result,

$$\int_{-\infty}^{\infty} f(x)\, dx = 2\pi i \cdot \frac{1}{4i} = \frac{\pi}{2}.$$

Example 5.12. The technique of example 5.11 is also useful in dealing with integrals of the form

$$\int_{-\infty}^{\infty} f(x) \cos ax\, dx \quad \text{or} \quad \int_{-\infty}^{\infty} f(x) \sin ax\, dx,$$

where a is a real number, and f is a rational function satisfying the same conditions as in 5.11.

Suppose for definiteness that $a \geqslant 0$. In this case we apply the residue theorem to $F(z) = f(z)\, e^{iaz}$, and notice that if $z \in \gamma_R^*$, $z = x + iy$, $y \geqslant 0$, and so

$$|F(z)| = |f(z)||e^{ia(x+iy)}| = |f(z)|e^{-ay} \leqslant |f(z)|.$$

Hence $\int_{\gamma_R} F(z)\,dz \to 0$ as before, and so

$$\int_{-\infty}^{\infty} F(z)\,dz = 2\pi i \sum_{k=1}^{\infty} R(F, a_k),$$

where a_k are the poles of f with $\operatorname{Im}(a_k) > 0$.

As an example of this technique, the reader should verify that for all real a,

$$\int_{-\infty}^{\infty} \frac{\cos ax}{1 + x^2}\,dx = \pi e^{-|a|},$$

a result which is of great importance in probability theory.

(Notice that it is not possible to apply the residue theorem to $f(z)\cos az$, since when $z = iy$ for real y, $\cos az = \cosh ay$ which tends rapidly to infinity with y.)

We now prove a technical result which enables us to deal with the case when $f(z)$ does not tend to zero as $|z|^{-2}$ when $|z| \to \infty$.

Lemma 5.13 (Jordan's Lemma). *Let f be continuous on the set $E = \{z\colon \operatorname{Im} z \geqslant 0\}$, and suppose $f(z) \to 0$ as $|z| \to \infty$ with $z \in E$. Let γ_R be the path given by $\gamma_R(t) = R\,e^{\pi i t}$, $0 \leqslant t \leqslant 1$, and let $a > 0$ be constant.*

Then $\int_{\gamma_R} f(z)\,e^{iaz}\,dz \to 0$ as $R \to \infty$.

Proof. Given $\varepsilon > 0$, let R_0 be such that $|f(z)| < \varepsilon$ if $|z| \geqslant R_0$. Then

$$\begin{aligned}\left|\int_{\gamma_R} f(z)\,e^{iaz}\,dz\right| &= \left|\int_0^1 f(R\,e^{i\pi t}) \exp\{iaR(\cos \pi t + i \sin \pi t)\}\, i\pi R\,e^{i\pi t}\,dt\right| \\ &\leqslant \varepsilon\pi R \int_0^1 e^{-aR\sin\pi t}\,dt = \varepsilon R \int_0^{\pi} e^{-aR\sin\theta}\,d\theta \\ &= 2\varepsilon R \int_0^{\pi/2} e^{-aR\sin\theta}\,d\theta.\end{aligned}$$

In order to estimate this integral, we make use of the geometrically obvious fact that for $0 \leqslant \theta \leqslant \frac{1}{2}\pi$, $2\theta/\pi \leqslant \sin\theta \leqslant \theta$. (An analytic proof of this may be based on the fact that

$$\frac{\sin\theta}{\theta} = \int_0^1 \cos(\theta t)\,dt$$

decreases from 1 to $2/\pi$ as θ increases from 0 to $\pi/2$.)

Then we have

$$2\varepsilon R \int_0^{\pi/2} e^{-aR\sin\theta}\, d\theta \leqslant 2\varepsilon R \int_0^{\pi/2} e^{-aR2\theta/\pi}\, d\theta = \frac{\pi\varepsilon}{a}(1 - e^{-aR}) < \frac{\pi\varepsilon}{a}.$$

Hence if $R \geqslant R_0$,

$$\left| \int_{\gamma_R} f(z)\, e^{iaz}\, dz \right| < \left(\frac{\pi}{a}\right) \varepsilon,$$

and the result follows.

As an application of Jordan's lemma, we prove that

$$\int_0^\infty \frac{\sin t}{t}\, dt,$$

an integral which converges conditionally, but not absolutely, has value $\frac{1}{2}\pi$.

Example 5.14. $\displaystyle\int_0^\infty \frac{\sin t}{t}\, dt = \tfrac{1}{2}\pi.$

Consider $f(z) = e^{iz}/z$. f is regular on $\mathbf{C}$ except for a simple pole at $z = 0$, with residue 1. Let $0 < r < R < \infty$, and take

$$\gamma = [-R, -r] \cup \gamma_r \cup [r, R] \cup \gamma_R,$$

where γ_R is as in 5.13, and

$$\gamma_r(t) = r\, e^{i\pi(1-t)}, \qquad 0 \leqslant t \leqslant 1.$$

Notice that since γ is disjoint from the negative half of the imaginary axis, the origin lies in the exterior of γ, and so $n(\gamma, 0) = 0$. It follows from the residue theorem, applied to $G = \mathbf{C}$, f and γ that $\int_\gamma f(z)\, dz = 0$.

We have that

$$\begin{aligned}\int_{[-R,-r]} f(z)\, dz + \int_{[r,R]} f(z)\, dz &= \int_{-R}^{-r} \frac{e^{ix}}{x}\, dx + \int_r^R \frac{e^{ix}}{x}\, dx \\ &= \int_r^R \frac{e^{ix} - e^{-ix}}{x}\, dx = 2i \int_r^R \frac{\sin x}{x}\, dx.\end{aligned}$$

In order to investigate the behaviour of $\int_{\gamma_r} f(z)\, dz$ as $r \to 0$, we write $f(z) = 1/z + g(z)$ where $g(z)$ is regular, and in particular is bounded, for $|z| \leqslant 1$: $|g(z)| \leqslant M$ if $|z| \leqslant 1$ say. Then

$$\int_{\gamma_r} f(z)\,dz = \int_{\gamma_r} \frac{dz}{z} + \int_{\gamma_r} g(z)\,dz.$$

The first term is

$$\int_0^1 -i\pi \frac{r\,e^{i\pi(1-t)}}{re^{i\pi(1-t)}}\,dt = -i\pi,$$

and the second term is bounded in modulus by $\pi r M$ (if $r \leqslant 1$) and so tends to zero with r.

An application of 5.13 shows that $\int_{\gamma_R} f(z)\,dz \to 0$ as $R \to \infty$.

Combining these results together, we have that

$$2i \int_r^R \frac{\sin x}{x}\,dx - i\pi \to 0 \quad \text{as} \quad r \to 0 \quad \text{and} \quad R \to \infty;$$

in other words the integral $\int_0^\infty [(\sin x)/x]\,dx$ converges, with value $\pi/2$.

Obvious changes of variable show that

$$\int_0^\infty \frac{\sin (ax)}{x}\,dx = \pm\frac{\pi}{2}$$

according as a is >0 or <0.

The technique used in 5.14 to deal with $\int_{\gamma_r} f(z)\,dz$ is sufficiently useful to be worth stating separately.

Lemma 5.15. *Let f have a simple pole at $a \in \mathbf{C}$, and let $\gamma_r(\theta) = a + r\,e^{i\theta}$, $\theta \in [\alpha, \beta]$. Then*

$$\int_{\gamma_r} f(z)\,dz \to i(\beta - \alpha)R(f, a) \quad \text{as} \quad r \to 0.$$

Proof. We have that $f(z) = R(f, a)(z - a)^{-1} + g(z)$, where $g(z)$ is bounded near a, say $|g(z)| \leqslant M$ if $|z - a| \leqslant r_0$.

Then if r $\leqslant r_0$,

$$\int_{\gamma_r} f(z)\,dz = R(f, a) \int_{\gamma_r} \frac{dz}{z - a} + \int_{\gamma_r} g(z)\,dz.$$

The first of these terms has value $i(\beta - \alpha)R(f, a)$, and the second is bounded in modulus by $M(\beta - \alpha)r$, and so tends to zero with r.

Our next use of the residue theorem is for the case of integrals involving fractional powers, or logarithms (of the real variable of integration). Since we have no corresponding complex functions (see Appendix C), we make some preliminary transformations to avoid this difficulty.

Example 5.16. For $0 < a < 1$,

$$\int_0^\infty \frac{x^{a-1}}{1+x}\,dx = \pi \operatorname{cosec}(a\pi).$$

We first make the substitution $x = e^t$, so that

$$\int_0^\infty \frac{x^{a-1}}{1+x}\,dx = \int_{-\infty}^\infty \frac{e^{at}}{1+e^t}\,dt.$$

Notice that the condition $a > 0$ is required to make the integral converge at the lower end of the range of integration ($x = 0$ or $t = -\infty$), while $a < 1$ ensures convergence at the upper end ($x = +\infty$ or $t = +\infty$).

We now take $f(z) = e^{az}/(1 + e^z)$, and notice that f has simple poles at $z = \pm i\pi, \pm 3i\pi, \pm 5i\pi, \ldots$. In order to exclude all but a finite number (in fact one) of these from consideration, we take for G the set $\{z: -\frac{1}{2}\pi < \operatorname{Im} z < 5\pi/2\}$. Then $i\pi$ is the only singularity of f in G; it is a simple pole, and

$$R(f, i\pi) = \left.\frac{e^{az}}{e^z}\right|_{z=i\pi} = -e^{ia\pi}.$$

Having chosen a band of finite width for G, we are prevented from using a semicircular path for γ, and choose instead the rectangle with vertices at R, $R + 2\pi i$, $-S + 2\pi i$, $-S$, where R, S will tend independently to infinity.

Since γ crosses the segment $[i\pi, -i\pi]$ in the positive direction, and $-i\pi \in E(\gamma)$, it follows from 5.7 that $n(\gamma, i\pi) = 1$.

The residue theorem now shows that

$$\int_\gamma f(z)\,dz = 2\pi i n(\gamma, i\pi)R(f, i\pi) = -2\pi i\, e^{ia\pi}.$$

The integral along the upper edge $[R + 2\pi i, -S + 2\pi i]$ of γ may be evaluated in terms of the integral along $[-S, R]$: in fact

$$\int_{[R+2\pi i, -S+2\pi i]} f(z)\,dz = -\int_{-S}^R f(x + 2\pi i)\,dx,$$

on writing $z = x + 2\pi i$. But

$$f(x + 2\pi i) = \frac{e^{a(x+2\pi i)}}{1 + e^{(x+2\pi i)}} = e^{2\pi a i} f(x),$$

and so

$$\int_{[-S,R]} f(z)\, dz + \int_{[R+2\pi i,\, -S+2\pi i]} f(z)\, dz = (1 - e^{2\pi a i}) \int_{-S}^{R} f(x)\, dx.$$

The integrals along the ends $[R, R + 2\pi i]$, $[-S + 2\pi i, -S]$ tend to zero, as is shown by the following considerations.

On $[R, R + 2\pi i]$, $z = R + iy$, $0 \leqslant y \leqslant 2\pi$, and so

$$\int_{[R,R+2\pi i]} f(z)\, dz = \int_0^{2\pi} \frac{e^{a(R+iy)} i}{1 + e^{(R+iy)}}\, dy,$$

which is bounded in modulus by $[e^{aR}/(e^R - 1)] \cdot 2\pi$ and tends to zero as $R \to \infty$, since $a < 1$.

Similarly it can be shown that

$$\left| \int_{[-S+2\pi i,\, -S]} f(z)\, dz \right| \leqslant \frac{e^{-aS} 2\pi}{1 - e^{-S}}$$

which tends to zero as $S \to \infty$, since $a > 0$.

Collecting these results, and letting R, S tend to infinity, we have that

$$(1 - e^{2\pi i a}) \int_{-\infty}^{\infty} \frac{e^{ax}}{1 + e^x}\, dx = -2\pi i\, e^{ia\pi}.$$

It follows that

$$\int_0^{\infty} \frac{x^{a-1}}{1 + x}\, dx = \int_{-\infty}^{\infty} \frac{e^{ax}}{1 + e^x}\, dx = \frac{\pi\, e^{ia\pi} 2i}{e^{2\pi i a} - 1}$$

$$= \frac{2i\pi}{e^{i\pi a} - e^{-i\pi a}} = \frac{\pi}{\sin a\pi}.$$

Corollary 5.17. *If $0 < a < b$, the substitution $y^b = x$ shows that*

$$\int_0^{\infty} \frac{y^{a-1}\, dy}{1 + y^b} = \frac{1}{b} \int_0^{\infty} \frac{x^{a/b - 1}}{1 + x}\, dx = \frac{\pi}{b \sin(\pi a/b)},$$

a result which contains many elementary results (with integer values of a, b) as special cases.

For a final example we use a combination of several of these techniques.

Example 5.18. $\int_0^\infty \frac{\log x}{x^2 - 1} dx = \frac{\pi^2}{4}.$

Notice that the integrand has only a logarithmic singularity as $x \to 0$; it tends to $\frac{1}{2}$ as $x \to 1$ (by l'Hopital's rule), and is bounded by $(2\sqrt{x})/(x^2 - 1)$ for $x > 1$. Hence the integral converges.

In order to use complex variable theory, we put $x = e^t$ and obtain

$$\int_0^\infty \frac{\log x}{x^2 - 1} dx = \int_{-\infty}^\infty \frac{t\, e^t}{e^{2t} - 1} dt.$$

We take $f(z) = z\, e^z/(e^{2z} - 1)$. f has a simple pole when $z = k\pi i$ for any integer $k \neq 0$, and a removable singularity at $z = 0$. For γ we choose the contour specified in figure 5.2; in particular

$$\gamma_r(t) = i\pi + r\, e^{-i\pi t}, \qquad 0 \leqslant t \leqslant 1.$$

The residue of f at $i\pi$ is

$$R(f, i\pi) = \left.\frac{z\, e^z}{2e^{2z}}\right|_{z=i\pi} = -\frac{i\pi}{2}.$$

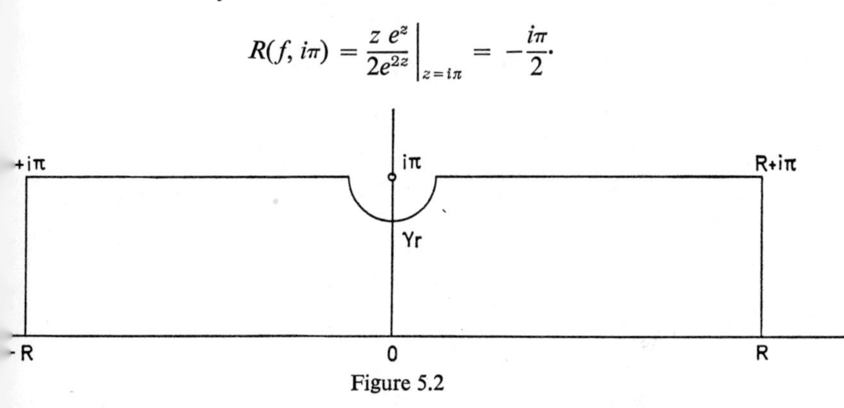

Figure 5.2

Finally we choose for G the strip $\{z: -\frac{1}{2}\pi < \operatorname{Im} z < 3\pi/2\}$, and apply the residue theorem to G, f and γ. The only singularity of f in G is at $i\pi$, which is in $E(\gamma)$, and so $\int_\gamma f(z)\, dz = 0$.

On the upper edge of the rectangle

$$f(x + i\pi) = \frac{(x + i\pi)e^{x+i\pi}}{e^{2x+2i\pi} - 1} = -\frac{(x + i\pi)\, e^x}{e^{2x} - 1}$$

so

$$f(x + i\pi) = -f(x) + i\frac{\pi\, e^x}{e^{2x} - 1}.$$

But $e^x/(e^{2x} - 1)$ is an odd function of x, so

$$\int_{-R}^{-r} + \int_{r}^{R} \frac{e^x}{e^{2x} - 1}\, dx = 0.$$

As $r \to 0$, it follows from lemma 5.15 (applied to $-\gamma_r$) that

$$\int_{\gamma_r} f(z)\, dz \to -i\pi\left(-\frac{i\pi}{2}\right) \quad \text{as} \quad r \to 0;$$

while as $R \to \infty$, a similar estimate to that in 5.16 shows that the integrals along $[R, R + i\pi]$ and $[-R + i\pi, -R]$ tend to zero as $R \to \infty$.

Combining these results, we find that

$$(1 - (-1)) \int_{-R}^{R} f(x)\, dx + i\pi \left(\frac{i\pi}{2}\right) \to 0 \quad \text{as} \quad R \to \infty.$$

Hence $\int_{-\infty}^{\infty} f(x)\, dx = \pi^2/4$ as required.

[It follows from integration by parts that for $r = 0, 1, 2, \ldots,$

$$\int_0^1 \log x\, x^r\, dx = -\frac{1}{(r + 1)^2}.$$

The reader who is familiar with Lebesgue's dominated convergence theorem will then be able to justify the steps in the following argument.

$$\int_0^{\infty} \frac{\log x}{x^2 - 1}\, dx = 2\int_0^1 \frac{\log x}{x^2 - 1}\, dx$$

$$= -2\int_0^1 \log x(1 + x^2 + x^4 + \cdots)\, dx = 2\left(1 + \frac{1}{3^2} + \frac{1}{5^2} + \cdots\right) = \frac{\pi^2}{4}$$

where the sum $\sum_{r=0}^{\infty} (2r + 1)^{-2}$ is a standard series whose sum may be found by the method of 5.20.]

We now turn to the use of the residue theorem to evaluate sums of the form $\sum_{n=-\infty}^{\infty} f(n)$, where f is a function which is regular on $\mathbf{C}$, except for a finite number of singularities which are disjoint from the integers. We exploit the fact (noted in 4.13) that the function $\pi \cot \pi z$ has a simple pole with residue 1 at each integer. Hence $F(z) = f(z)\pi \cot \pi z$ has a simple pole at $z = n$, $R(F, n) = f(n)$. If $\{a_1, a_2, \ldots, a_k\}$ are the singularities of f, and S_n is the square with vertices $(n + \frac{1}{2})(\pm 1 \pm i)$, as in example 5.8, then we may apply the residue theorem to F on the domain $\{z: |\text{Re } z| < n + 1\}$ to obtain

$$\int_{S_n} F(z)\, dz = 2\pi i \left\{ \sum_{m=-n}^{n} f(m) + \sum_{j=1}^{k} R(F, a_j) \right\},$$

provided $n > \max \{|a_j|; 1 \leqslant j \leqslant k\}$. To proceed further we need a technical result which determines the size of $\cot \pi z$ for $z \in S_n^*$.

Lemma 5.19 (i) *If* x, y *are real, then* $|\cos (x + iy)|$ *and* $|\sin (x + iy)|$ *both lie between* $\sinh |y|$ *and* $\cosh y$.

(ii) *If* S_n *is the square described above, then* $|\cot \pi z| \leqslant 2$ *for all* $z \in S_n^*$, $n = 0, 1, 2, \ldots$.

Proof. (i) We prove the result for $\cos (x + iy)$; that for $\sin (x + iy)$ is similar.

We have from the results of 1.8 that

$$\cos (x + iy) = \cos x \cosh y - i \sin x \sinh y,$$

and hence that

$$\begin{aligned} |\cos (x + iy)|^2 &= \cos^2 x \cosh^2 y + \sin^2 x \sinh^2 y \\ &= \cosh^2 y + \sin^2 x(\sinh^2 y - \cosh^2 y) = \cosh^2 y - \sin^2 x. \end{aligned}$$

Hence $\sinh^2 y = \cosh^2 y - 1 \leqslant |\cos (x + iy)|^2 \leqslant \cosh^2 y$, from which the result follows.

(ii) Suppose $z \in S_n^*$. Then there are two possibilities

(a) $z = \pm(n + \frac{1}{2}) + iy, \quad |y| \leqslant (n + \frac{1}{2})$.

In this case

$$|\cot \pi z| = \left| \frac{\pm \sin (n + \frac{1}{2})\pi \sinh y\pi}{\pm \sin (n + \frac{1}{2})\pi \cosh y\pi} \right| = |\tanh (y\pi)| < 1.$$

(b) $z = x + i(n + \frac{1}{2}), \quad |x| \leqslant (n + \frac{1}{2})$.

Then it follows from (i) that

$$\begin{aligned} |\cot \pi z| &\leqslant \frac{\cosh (n + \frac{1}{2})\pi}{\sinh (n + \frac{1}{2})\pi} \leqslant \coth (\tfrac{1}{2}\pi) < \coth (1.5) \\ &= 1.105\ldots < 2. \end{aligned}$$

We can now prove our main result on the summation of series.

Theorem 5.20. *Let f be a function which is regular on* $\mathbf{C}$ *except for singularities at a finite number of points* $a_1, a_2, \ldots, a_k$, *where* $a_j \notin \mathbf{Z}$, $j = 1, 2, \ldots, k$. *Suppose also that* $|f(z)| \leqslant M|z|^{-2}$ *for sufficiently large* $|z|$.

Then

$$\sum_{n=-\infty}^{\infty} f(n) = -\sum_{j=1}^{k} R(F, a_j), \quad \text{where} \quad F(z) = f(z)\pi \cot \pi z.$$

(Notice that the conditions are satisfied in particular if f is a rational function, whose poles are disjoint from the integers, and which has a zero of order at least two at infinity.)

Proof. The discussion preceding 5.19 shows that if

$$F(z) = f(z)\pi \cot \pi z,$$

then

$$\sum_{m=-n}^{n} f(m) = -\sum_{j=1}^{k} R(F, a_j) + \frac{1}{2\pi i}\int_{S_n} F(z)\, dz,$$

provided

$$n > \max\{|a_j|, 1 \leqslant j \leqslant k\}.$$

Suppose that $|f(z)| \leqslant M|z|^{-2}$ if $|z| \geqslant R_0$.

Then for $n \geqslant R_0$,

$$\left|\frac{1}{2\pi i}\int_{S_n} f(z)\pi \cot \pi z\, dz\right| \leqslant \frac{M}{2\pi} L(S_n) n^{-2}\, 2\pi$$

using 1.20 (vi) and (ii) of 5.19.

But $L(S_n) = 4(2n + 1)$ and so

$$\left|\frac{1}{2\pi i}\int_{S_n} F(z)\, dz\right| \leqslant 4\frac{M(2n+1)}{n^2}$$

which tends to zero as $n \to \infty$, and the result follows.

Example 5.21. Let a be a complex number which is not an integer. Then

$$\frac{1}{a} + \sum_{k=1}^{\infty} \frac{2a}{a^2 - k^2} = \pi \cot \pi a.$$

We apply theorem 5.20 with

$$f(z) = \frac{2a}{a^2 - z^2}, \qquad F(z) = f(z)\pi \cot \pi z.$$

Then F has simple poles at $z = \pm a$,

$$R(F, \pm a) = \frac{2a\pi \cot (\pm a\pi)}{-2(\pm a)} = -\pi \cot a\pi$$

in each case. It follows that

$$\sum_{m=-\infty}^{\infty} \frac{2a}{a^2 - m^2} = -2(-\pi \cot a\pi)$$

for all $a \notin \mathbf{Z}$.

If we group the terms with $\pm m = k$, $k = 1, 2, \ldots$, then we obtain

$$\frac{1}{a} + 2 \sum_{k=1}^{\infty} \frac{a}{a^2 - k^2} = \pi \cot \pi a, \quad \text{as required.}$$

Since the convergence is uniform on any compact set disjoint from $\mathbf{Z}$, we may apply theorem 3.11 to differentiate term by term to obtain

$$-\frac{1}{a^2} + \sum_{k=1}^{\infty} \frac{d}{da}\left(\frac{1}{a-k} + \frac{1}{a+k}\right) = -\pi^2 \operatorname{cosec}^2 \pi a,$$

or

$$\pi^2 \operatorname{cosec}^2 \pi a = \sum_{k=-\infty}^{\infty} \frac{1}{(a-k)^2}.$$

§3. ROUCHÉ'S THEOREM AND THE LOCAL MAPPING THEOREM

We now turn to an application of the residue theorem which gives information concerning the number of zeros and poles of a regular function in a starred open set.

Definition 5.22. Let f be regular, or have a pole at the point $a \in \mathbf{C}$. We define the order of f at a,

$$\begin{aligned} O(f, a) &= +k, && \text{if } f \text{ has a zero of order } k \text{ at } a,\ k = 1, 2, \ldots, \\ &= 0, && \text{if } f \text{ is regular, but non-zero at } a, \\ &= -k, && \text{if } f \text{ has a pole of order } k \text{ at } a,\ k = 1, 2, \ldots. \end{aligned}$$

Notice that in each case, if $r = O(f, a)$, then $g(z) = (z - a)^{-r}f(z)$ is regular and non-zero at a.

Theorem 5.23. *Let G be a starred open set, and $A = \{a_1, a_2, \ldots, a_n\}$, a finite subset of it.*

Let f be regular and non-zero on $G \setminus A$, and have a zero, or a pole at each point of A. Let γ be a closed path, with $\gamma^* \subset G \setminus A$.

Then

$$\frac{1}{2\pi i}\int_\gamma \frac{f'(z)}{f(z)}\,dz = \sum_{j=1}^{n} O(f, a_j)n(\gamma, a_j).$$

Proof. Write $r_j = O(f, a_j)$: as noted in 5.22, the function

$$(z - a_j)^{-r_j}f(z)$$

is regular and non-zero at a_j.

Hence

$$g(z) = \prod_{j=1}^{n} (z - a_j)^{-r_j}f(z)$$

is regular and never zero on G.

It follows that $g'(z)/g(z)$ is regular on G, and so by Cauchy's theorem (2.3), that

$$\int_\gamma \frac{g'(z)}{g(z)}\,dz = 0.$$

However,

$$\frac{g'(z)}{g(z)} = \frac{f'(z)}{f(z)} - \sum_{j=1}^{n} \frac{r_j}{z - a_j},$$

and so

$$\frac{1}{2\pi i}\int_\gamma \frac{f'(z)}{f(z)}\,dz = \frac{1}{2\pi i}\int_\gamma \frac{g'(z)}{g(z)}\,dz + \sum_{j=1}^{n} r_j \frac{1}{2\pi i}\int_\gamma \frac{dz}{z - a_j}$$

$$= 0 + \sum_{j=1}^{n} r_j n(\gamma, a_j),$$

as required.

Corollary 5.24. *Let G, A and f be as in 5.23, and γ a closed path, with $\gamma^* \subset G \setminus A$ and having a single bounded component B, on which*

$n(\gamma, w) = 1$. (Examples of suitable paths are circles (1.22), squares (5.8) and semi-circles (5.8); other examples may be identified by means of theorem 5.7.)

Then

$$\frac{1}{2\pi i}\int_\gamma \frac{f'(z)}{f(z)}\,dz = \sum_{a_j\in B} O(f, a_j).$$

[This result is often paraphrased by saying that the value of the integral of f'/f along γ is the sum of the orders of f at the points in the bounded component of γ. More simply it is the number of zeros, minus the number of poles, each counted according to multiplicity.]

Proof. According to 5.22, we have

$$\frac{1}{2\pi i}\int_\gamma \frac{f'(z)}{f(z)}\,dz = \sum_{j=1}^{n} O(f, a_j)n(\gamma, a_j).$$

But by our hypothesis on γ, $n(\gamma, a_j) = 1$ if $a_j \in B$, while $n(\gamma, a_j) = 0$ for the other values of j, and the result follows.

Corollary 5.24 is most often used when f has only zeros in G: in this case if γ and B are as before, $(1/2\pi i)\int_\gamma [f'(z)/f(z)]\,dz$ gives the number of zeros of f in B. Another way of viewing this result is as follows.

Let $\Gamma(t) = f(\gamma(t))$: Γ is the curve described by $f(z)$ when z describes γ. Then writing $w = f(z)$,

$$\frac{1}{2\pi i}\int_\gamma \frac{f'(z)}{f(z)}\,dz = \frac{1}{2\pi i}\int_\Gamma \frac{dw}{w} = n(\Gamma, 0),$$

the index of zero with respect to Γ.

Similarly if a is any other complex number, the number of times $f(z) = a$ in B (counted according to multiplicity) is equal to

$$\frac{1}{2\pi i}\int_\gamma \frac{f'(z)\,dz}{f(z) - a} = \frac{1}{2\pi i}\int_\Gamma \frac{dw}{w - a} = n(\Gamma, a).$$

A consequence of theorem 5.23 is that if f, g are functions whose values are fairly 'close' on path γ, then f and g have the same number of zeros in the bounded component of γ. These ideas are made precise in the following result.

Theorem 5.25 (Rouché). *Let G be a starred open set, and f, g be regular functions on G.*

Let γ be a closed path with $\gamma^* \subset G$, and having a single bounded component B with $n(\gamma, w) = 1$ for $w \in B$.

Suppose that for each $z \in \gamma^*$,

$$|f(z) - g(z)| < |f(z)| + |g(z)|. \qquad (*)$$

Then f and g have the same number of zeros in B.

Proof. Notice that the condition (*) implies that neither f nor g can be zero on γ^*.

Notice also that for any complex numbers w, w',

$$|w - w'| \leqslant |w| + |w'|,$$

and that if w, w' are non-zero, equality is possible only if w/w' is real and strictly negative. Hence our condition (*) says that in addition to the non-vanishing of f and g on γ^*, $f(z)/g(z)$ is never real and negative if $z \in \gamma^*$.

We now consider the linear combination

$$f_\lambda(z) = (1 - \lambda)f(z) + \lambda g(z)$$

for $0 \leqslant \lambda \leqslant 1$. Firstly we have $f_0 = f$, and $f_1 = g$; and secondly if $0 < \lambda < 1$, $f_\lambda(z) = 0$ implies $f(z)/g(z) = \lambda/(\lambda - 1)$ which is real and <0, and we know this to be impossible. It follows that for $0 \leqslant \lambda \leqslant 1$, $f_\lambda(z) \neq 0$ if $z \in \gamma^*$.

We can now define for each $\lambda \in [0, 1]$ a function

$$p(\lambda) = \frac{1}{2\pi i}\int_\gamma \frac{f'_\lambda(z)}{f_\lambda(z)}\,dz = \frac{1}{2\pi i}\int_\gamma \frac{(1 - \lambda)f'(z) + \lambda g'(z)}{(1 - \lambda)f(z) + \lambda g(z)}\,dz,$$

which gives (by corollary 5.24) the number of zeros of f_λ in B, and so is integer valued. But p is certainly a continuous function of $\lambda \in [0, 1]$ and hence must be constant.

In particular $f_0 = f$ and $f_1 = g$ have the same number of zeros in B, which completes the proof.

Rouché's theorem is used to locate zeros of a function by comparing it with another function whose zeros may be located more easily, as in the following example.

Example 5.26. Show that for $z \in S(0, 1)$, the equation

$$z\, e^{\lambda - z} = 1$$

where λ is real and >1, has a unique solution, which is real and >0.

We consider $f(z) = z - e^{z-\lambda}$. $f(z)$ is real valued if z is real, and $f(0) < 0$, $f(1) = 1 - e^{1-\lambda}$, which is positive since $\lambda > 1$. Hence there is (at least) one real zero in the interval $(0, 1)$, and we use Rouché's theorem to show that this is the only zero in $S(0, 1)$.

Let $g(z) = z$: evidently g has a simple zero at the origin.

We take $\gamma(t) = e^{2\pi i t}$, $0 \leqslant t \leqslant 1$, when $\gamma^* = C(0, 1)$. Then if $z \in \gamma^*$, $|z| = 1$, and

$$\begin{aligned} |f(z) - g(z)| &= |e^{z-\lambda}| = \exp(\operatorname{Re} z - \lambda) \\ &\leqslant \exp(1 - \lambda) < 1 = |g(z)| \leqslant |f(z)| + |g(z)|. \end{aligned}$$

Hence Rouché's theorem may be applied, and the result follows.

Rouché's theorem may also be used to investigate the behaviour of a regular function in the neighbourhood of an n-fold zero, as the following theorem demonstrates.

Theorem 5.27. *Let f be regular and non-constant in an open set G, $z_0 \in G$, $f(z_0) = w_0$, and suppose that the order of the zero of $f(z) - f(z_0)$ at $z = z_0$ is n $(n = 1, 2, 3, \ldots)$.*

Then there is an $r_0 > 0$, such that for any ε, $0 < \varepsilon \leqslant r_0$, there is a $\delta > 0$, depending on ε such that if $0 < |w - w_0| < \delta$, there are n distinct values of z with $0 < |z - z_0| < \varepsilon$, for which $f(z) = w$.

Proof. Since f is non-constant, the zero of $f(z) - f(z_0)$ at z_0 is isolated and there is $r_1 > 0$, for which $f(z) \neq f(z_0)$ if

$$0 < |z - z_0| \leqslant r_1.$$

If $n = 1, f'(z_0) \neq 0$, while if $n \geqslant 2, f'(z_0) = 0$ and f' is non-constant; in either case there is $r_2 > 0$ for which $f'(z) \neq 0$ if

$$0 < |z - z_0| \leqslant r_2.$$

Take $r_0 = \min(r_1, r_2)$ and let ε, $0 < \varepsilon \leqslant r_0$ be given.

The circle $C(z_0, \varepsilon)$ is a compact set, and by our construction $|f(z) - f(z_0)|$ is a continuous positive function which does not vanish on it. It follows from 1.4 that

$$\delta = \inf\{|f(z) - f(z_0)|;\, z \in C(z_0, \varepsilon)\} > 0.$$

We show that this δ has the required properties.

Suppose that $0 < |w - w_0| < \delta$.

Then if $z \in C(z_0, \varepsilon)$,

$$|(f(z) - w) - (f(z) - w_0)| = |w - w_0| < \delta \leqslant |f(z) - f(z_0)|$$
$$= |f(z) - w_0| \leqslant |f(z) - w_0| + |f(z) - w|.$$

It follows from Rouché's theorem that $f(z) - w$ and $f(z) - w_0$ have the same number of zeros in $S(z_0, \varepsilon)$. However, we know that $f(z) - w_0$ has just an n-fold zero at z_0, and so $f(z) - w$ has n zeros with $0 < |z - z_0| < \varepsilon$. For such values, we have ensured that $f'(z)$ does not vanish, and hence the zeros are necessarily simple.

Corollary 5.28. *Let G be an open set and f a non-constant regular function on G.*

Then (i) $f(G)$ is open, and

(ii) if in addition f is one-to-one, so that the inverse function f^{-1} exists, f^{-1} is also continuous (see also 5.33).

Proof. (i) The above theorem shows that if $w \in f(G)$, there is $\delta > 0$ for which $S(w, \delta) \subset f(G)$. In other words, $f(G)$ is an open set.

(ii) Let f be one-to-one, $z_0 \in G, f(z_0) = w_0$, and $\varepsilon > 0$ be given.

Then if $\varepsilon' = \min(\varepsilon, r_0)$ where r_0 is given by 5.27, there is $\delta > 0$, such that if $|w - w_0| < \delta$, the (necessarily unique) z with $f(z) = w$ satisfies $|z - z_0| < \varepsilon$.

In other words

$$|f^{-1}(w) - f^{-1}(w_0)| < \varepsilon \quad \text{if} \quad |w - w_0| < \delta,$$

and f^{-1} is continuous.

We finish this chapter with two results, one of which (5.29) is a new proof of a result given in an exercise to Chapter 3, and the other (5.33) clarifies further the relations between a one-to-one regular function and its inverse.

Theorem 5.29 (Maximum modulus theorem). *Let f be regular and non-constant on an open set G. Then $|f|$ does not have a local maximum at any point of G.*

Proof. Suppose that $|f|$ had a local maximum at $z_0 \in G$.

Then for some $r > 0$, $|f(z)| \leqslant |f(z_0)|$ for all z with $|z - z_0| \leqslant r$.

But for any $\delta > 0$, $S(f(z_0), \delta)$ contains points with modulus greater than $|f(z_0)|$, and hence f cannot map onto $S(f(z_0), \delta)$, and 5.25 is contradicted.

Corollary 5.30. *Let f be regular and not constant on an open set G, and K be a compact subset of G. Suppose that $|f(z)| \leqslant M$ for $z \in K$, and that $S(z_0, r) \subset K$* (in the terminology of 1.2, z_0 is in the interior of *K). Then $|f(z_0)| < M$.*

Proof. Since K is compact, and $|f|$ is continuous on K, $|f|$ must attain a maximum relative to K at a point $p \in K$, $|f(z)| \leqslant |f(p)| \leqslant M$ for all $z \in K$.

If $S(z_0, r) \subset K$, and $|f(z_0)| = M$, then $|f|$ must have a local maximum at z_0, contrary to 5.29. Hence $|f(z_0)| < M$.

Definition 5.31. A regular function on an open set G is called univalent if it is one-to-one in G. (The German word schlicht is also sometimes used.)

Lemma 5.32. *Let f be univalent in an open set G. Then $f'(z) \neq 0$ for any $z \in G$.*

Proof. If $f'(z_0) = 0$ for some $z_0 \in G$, then $f(z) - f(z_0)$ has a zero of order at least two at z_0. Hence by 5.27, f must be n-to-one in a neighbourhood of z_0, for some $n \geqslant 2$, and so not univalent.

It should be noticed that the converse of 5.32 holds locally by theorem 5.27; if $f'(z_0) \neq 0$, then f is one-to-one in some $S(z_0, r)$. However, the converse does not hold in larger regions—for instance if $f(z) = e^z$ for $z \in \mathbf{C}$, $f'(z) = e^z$ is never zero, but $f(z) = f(z + 2\pi i)$ for all $z \in \mathbf{C}$.

Theorem 5.33. *Let f be regular and univalent on the open set G. By 5.28, $G_1 = f(G)$ is open and f^{-1} is continuous on G_1. Then in addition $g = f^{-1}$ is regular on G_1, and if $z_0 \in G, f(z_0) = w_0 \in G_1$, then $g'(w_0) = 1/f'(z_0)$.*

Proof. Let $w, w_0 \in G_1, f(z) = w, f(z_0) = w_0$.
Then

$$\frac{g(w) - g(w_0)}{w - w_0} = \frac{f^{-1}(w) - f^{-1}(w_0)}{w - w_0} = \frac{z - z_0}{f(z) - f(z_0)}.$$

We know already that f^{-1} is continuous; that is

$$f^{-1}(w) = z \to z_0 = f^{-1}(w_0) \quad \text{as} \quad w \to w_0.$$

Hence

$$\lim_{w \to w_0} \frac{g(w) - g(w_0)}{w - w_0} = \lim_{z \to z_0} \frac{z - z_0}{f(z) - f(z_0)},$$

and since $f'(z_0) \neq 0$ by 5.32, the right-hand limit exists, and its value is $1/f'(z_0)$.

EXERCISES FOR CHAPTER 5

1. Consider the following closed paths in $\mathbf{C}$:

(i) $\gamma(t) = \begin{cases} -1 + 4t, & 0 \leqslant t \leqslant \frac{1}{2}, \\ \exp\{2\pi i(\frac{1}{2} - t)\}, & \frac{1}{2} < t < 1. \end{cases}$

(ii) $\gamma(t) = \begin{cases} t + 2it^2, & 0 \leqslant t \leqslant \frac{1}{2}, \\ (1 + i)(1 - t), & \frac{1}{2} < t < 1. \end{cases}$

(iii) $\gamma(t) = \sin^2(\pi t)\, e^{2\pi i t}, \quad 0 \leqslant t \leqslant 1.$

(iv) $\gamma(t) = e^{4\pi i t}, \quad 0 \leqslant t \leqslant 1.$

(v) $\gamma(t) = \begin{cases} t + i\sin(4\pi t), & 0 \leqslant t \leqslant \frac{1}{2}, \\ 1 - t, & \frac{1}{2} < t < 1. \end{cases}$

In each case, determine the components of $\mathbf{C} \setminus \gamma^*$, and find the value of the topological index on each component.

2. Is it possible for a closed path to have a single bounded component on which the topological index is zero?

3. Show how the integral formulae for the derivatives of a regular function (theorem 2.7) may be regarded as a special case of the residue theorem.

4. Evaluate the following integrals by the technique of example 5.10.

(i) $\displaystyle\int_0^{2\pi} \frac{\sin^2\theta}{2 + \cos\theta}\, d\theta$, (ii) $\displaystyle\int_0^{\pi} \frac{d\theta}{1 + \cos^2\theta}$,

(iii) $\displaystyle\int_0^{\pi} \frac{\theta}{1 + \cos^2\theta}\, d\theta$ (put $\pi - \theta$ for θ).

5. Evaluate the following integrals by the technique of example 5.11.

$$\text{(i)} \int_{-\infty}^{\infty} \frac{dx}{1+x+x^2}, \quad \int_{-\infty}^{\infty} \frac{dx}{1+x^4}, \quad \int_{-\infty}^{\infty} \frac{x^2\,dx}{1+x^4}.$$

6. Evaluate the following integrals by the technique of examples 5.12–5.15.

$$\text{(i)} \int_{-\infty}^{\infty} \frac{x \sin x}{x^2+4}\,dx, \quad \text{(ii)} \int_{0}^{\infty} \frac{\cos x\,dx}{1+x^4}, \quad \text{(iii)} \int_{-\infty}^{\infty} \frac{\sin \pi x}{x^2-1}\,dx,$$

$$\text{(iv)} \int_{0}^{\infty} \frac{\sin^2 x}{x^2}\,dx.$$

7. Evaluate the following integrals by the techniques of examples 5.16 and 5.18.

$$\text{(i)} \int_{0}^{\infty} \frac{\sqrt{x}}{(1+x)^2\,dx}, \quad \text{(ii)} \int_{0}^{\infty} \frac{\log x}{1+x^2}\,dx, \quad \text{(iii)} \int_{1}^{\infty} \frac{dx}{x\sqrt{(x^2-1)}}$$

(put $x = \cosh t$).

8. Evaluate the following sums by the use of theorem 5.20 (or deduce their values from other results established in 5.21).

$$\text{(i)} \sum_{n=0}^{\infty} \frac{1}{n^2+1} = \tfrac{1}{2}\left(1 + \sum_{n=-\infty}^{\infty} \frac{1}{n^2+1}\right), \quad \text{(ii)} \sum_{n=0}^{\infty} \frac{1}{(2n+1)^2},$$

$$\text{(iii)} \sum_{n=-\infty}^{\infty} \frac{1}{(2n+1)(3n+1)}.$$

9. (Miscellaneous.) Establish the following results:

$$\text{(i)} \int_{-\infty}^{\infty} \frac{\cos ax - \cos bx}{x^2}\,dx = \pi(b-a),$$

$$\text{(ii)} \int_{-\infty}^{\infty} \frac{\sinh(cx)}{\sinh(\pi x)}\,dx = \tan(\tfrac{1}{2}c), \quad |c| < \pi,$$

$$\text{(iii)} \int_{0}^{\pi/2} \cos^{2n}\theta\,d\theta = \int_{0}^{\pi/2} \sin^{2n}\theta\,d\theta = \frac{\pi(2n)!}{2^{2n+1}(n!)^2},$$

$n = 0, 1, 2, \ldots.$

10. The integral $\int_0^\pi \log \sin\theta\,d\theta$ is important in applications, and is sometimes evaluated by complex methods.

Justify the steps in the following elementary real variable method:

$$I = \int_0^\pi \log \sin\theta\,d\theta = 2\int_0^{\pi/2} \log \sin\theta\,d\theta = 2\int_0^{\pi/2} \log \cos\theta\,d\theta.$$

Also

$$I = \int_0^\pi \log (2 \sin \tfrac{1}{2}\theta \cos \tfrac{1}{2}\theta)\, d\theta$$

$$= \int_0^\pi (\log 2 + \log \sin \tfrac{1}{2}\theta + \log \cos \tfrac{1}{2}\theta)\, d\theta$$

$$= \pi \log 2 + 2I.$$

Hence $I = -\pi \log 2$.

11. We assume the result that

$$I = \int_0^\infty e^{-t^2}\, dt = \tfrac{1}{2}\sqrt{\pi}.$$

[The traditional proof consists of justifying the following argument:

$$I^2 = \lim_{A\to\infty} \int_0^A \int_0^A e^{-(x^2+y^2)}\, dx\, dy = \lim_{R\to\infty} \int_0^R \int_0^{\pi/2} e^{-r^2}\, r\, d\theta\, dr$$

$$= \frac{\pi}{2} \int_0^\infty e^{-r^2}\, r\, dr = \frac{\pi}{4}.\Big]$$

Show, by integrating $f(z) = e^{-z^2/2}$ around the perimeter of a rectangle with vertices at R, $R + iu$, $-R + iu$, $-R$, and letting $R \to \infty$, that

$$\int_{-\infty}^\infty e^{-(x^2/2) - iux}\, dx = \sqrt{(2\pi)}\, e^{-u^2/2}.$$

12. Let $f(z) = z^n + a_{n-1}z^{n+1} + \cdots + a_1 z + a_0$, where $n \geqslant 1$, and $a_0, a_1, \ldots, a_{n-1}$ are complex numbers.

Let R_0 be the largest positive root of the equation

$$R^n = |a_{n-1}|R^{n+1} + \cdots + |a_1|R + |a_0|.$$

Prove that all the roots of $f(z) = 0$ satisfy $|z| \leqslant R_0$.
Hint: Let $g(z) = z^n$, estimate $|f - g|$ and $|f| + |g|$, and apply Rouché's theorem to $C(0, R)$ for any $R > R_0$.

13. Let λ be a real number >1, and let $f(z) = \lambda - z - e^{-z}$ for $z \in \mathbf{C}$. Show that in the halfplane $\{z\colon \operatorname{Re} z \geqslant 0\}$, f has a single zero which is real and positive.
Hint: Let $g(z) = \lambda - z$, and apply Rouché's theorem to a semicircular path with radius $R > \lambda$.

14. For each of the following functions, find a neighbourhood of the origin on which the conclusions of theorem 5.27 hold.

(i) $f(z) = z^2 + z^3$, $\quad z \in \mathbf{C}$,

(ii) $f(z) = \tan z, \quad z \in \mathbf{C} \setminus \mathbf{Z}$,
(iii) $f(z) = \sin z/z, \quad z \in \mathbf{C}$.

15. (Schwarz' lemma.) Let f be regular on $S(0, 1)$, and satisfy $|f(z)| \leqslant 1$ there. Suppose also that $f(0) = 0$. Show that $|f(z)| \leqslant |z|$ for all $z \in S(0, 1)$, and that there is equality at any $z \in S(0, 1)$ if and only if $f(z) = \alpha z$, for some complex number α, with $|\alpha| = 1$.
Hint: Apply the maximum modulus theorem to $g(z) = f(z)/z$ on $S(0, 1 - \varepsilon)$ where $\varepsilon > 0$.

16. Let f be regular on $G \supset \bar{S}(a, R)$, and suppose $|f|$ is constant on $C(a, R)$. Show that either f is constant on $\bar{S}(a, R)$, or f has at least one zero on $S(a, R)$.

6

Harmonic Functions and the Dirichlet Problem

In this chapter we consider the properties of the real and imaginary parts of a regular function, considered as functions in their own right. In particular we shall see that they are characterized by satisfying a certain mean value property, and also by being solutions of Laplace's equation

$$\Delta u = \frac{\partial^2 u}{\partial x^2} + \frac{\partial^2 u}{\partial y^2} = 0.$$

This fact is responsible for the great importance which such functions have in mathematical physics. We also treat a special case of the general problem of finding a harmonic function which has certain prescribed values on the boundary of its domain of definition. The second part of the chapter considers the relation between a harmonic function u and a conjugate function v for which $u + iv$ is regular.

§1. HARMONIC FUNCTIONS

We first show that the real and imaginary parts of a regular function have the properties mentioned above. This lemma should be compared with theorem 6.8 below.

Lemma 6.1. *Let G be open in* **C**, *and f be regular on G. Let*

$$u(x, y) = \operatorname{Re} f(x + iy), \quad \text{and} \quad v(x, y) = \operatorname{Im} f(x + iy)$$

be the real and imaginary parts of f, defined for $x + iy \in G$. We shall often write $u(z)$ in place of $u(x, y)$ when $z = x + iy$.

Then (i) u and v have partial derivatives of all orders and are solutions of Laplace's equations, and

(ii) if $\bar{S}(a, R) \subset G$, $u(a) = [1/(2\pi)] \int_0^{2\pi} u(a + R\, e^{i\theta})\, d\theta$, and the same property is valid for v also.

Proof. (i) We found in 1.11 that

$$\frac{\partial u}{\partial x}(x, y) = \text{Re} f'(x + iy) = \frac{\partial v}{\partial y}(x, y),$$

and

$$\frac{\partial u}{\partial y}(x, y) = -\text{Im} f'(x + iy) = -\frac{\partial v}{\partial x}(x, y).$$

From 2.7, f has derivatives of all orders, and hence u and v must have partial derivatives (necessarily continuous) of all orders: in particular the value of the mixed partial derivatives will be independent of the order in which the differentiations are performed.

Hence

$$\frac{\partial^2 u}{\partial x^2} = \frac{\partial}{\partial x}\left(\frac{\partial v}{\partial y}\right) = \frac{\partial}{\partial y}\left(\frac{\partial x}{\partial v}\right) = \frac{\partial}{\partial y}\left(-\frac{\partial u}{\partial y}\right) = -\frac{\partial^2 u}{\partial y^2},$$

so that u is a solution of Laplace's equation. A similar argument applies to v.

To prove (ii) we make use of the Cauchy integral formula, 2.5. This shows that if $\bar{S}(a, R) \subset G$,

$$f(a) = \frac{1}{2\pi i} \int_{C(a,R)} \frac{f(z)\, dz}{z - a}.$$

We now write $z = a + R\, e^{i\theta}$, $0 \leqslant \theta \leqslant 2\pi$, and obtain

$$f(a) = \frac{1}{2\pi} \int_0^{2\pi} f(a + R\, e^{i\theta})\, d\theta.$$

The required results follow on taking real and imaginary parts.

Definition 6.2. Let G be an open set in **C**, and u be a continuous real valued function on G. We say u is harmonic on G if for each $\bar{S}(a, R) \subset G$,

$$u(a) = \frac{1}{2\pi} \int_0^{2\pi} u(a + R\, e^{i\theta})\, d\theta.$$

If the sign of equality is replaced by $\leqslant$, then we say u is subharmonic on G.

Our objectives are to show that this property is equivalent to that of being a solution of Laplace's equation, and also (subject to certain conditions on G) that any harmonic function is the real part of a regular function.

Lemma 6.3. (i) *Let u, v be harmonic functions on an open set G. Then for any real constants a, b, $au + bv$ is also harmonic in G.*

(ii) *Suppose u, v are continuous on $\bar{S}(a, R)$, harmonic on $S(a, R)$ and $u = v$ at each point of $C(a, R)$. Then $u = v$ on $S(a, R)$ also.*

Proof. (i) is immediate, since the defining condition in 6.2 is linear in u.

(ii) By (i), $u - v$ is harmonic on $S(a, R)$. Suppose there is a point of $S(a, R)$ at which $u \neq v$—we may suppose $u > v$ without loss of generality.

Then

$$\alpha = \sup \{u(x, y) - v(x, y);\, x + iy \in S(a, R)\}$$

is strictly positive and is attained at some point $b \in S(a, R)$. We consider $r > 0$ with $\bar{S}(b, r) \subset S(a, R)$. Since $u - v = 0$ on $C(a, R)$, we may take r sufficiently close to $R - |b - a|$ to ensure that $u - v < \alpha$ at (at least) one point of $C(b, r)$, while the definition of α shows that $u - v \leqslant \alpha$ at all points of $C(b, r)$. The continuity of $u - v$ now shows that

$$\alpha = u(b) - v(b) > \frac{1}{2\pi} \int_0^{2\pi} \{u(b + r\, e^{i\theta}) - v(b + r\, e^{i\theta})\}\, d\theta,$$

which contradicts the harmonic property of $u - v$.

We now introduce the formula of Poisson which shows that, like the regular functions considered in Chapter 2, the values of a harmonic function at a point inside a circle are determined explicitly via an integral in terms of the values on the circumference.

Definition 6.4. The function

$$P_{R,r}(\theta) = \frac{R^2 - r^2}{R^2 - 2Rr \cos \theta + r^2},$$

defined for $0 < r < R$, and all real θ, is called the Poisson kernel. In the case $R = 1$, we shall write $P_r(\theta)$ for $P_{1,r}(\theta)$.

The function $P(g)$ defined in 6.6 below is called the Poisson integral of G.

Lemma 6.5. *As a function $z = r\,e^{i\theta}$, $P_{R,r}(\theta)$ is harmonic in $S(0, R)$ and has the properties*

(a) $\dfrac{1}{2\pi}\displaystyle\int_0^{2\pi} P_{R,r}(\theta)\,d\theta = 1$, and

(b) *for each δ, $0 < \delta < \pi$, $P_{R,r}(\theta) \to 0$ uniformly for $\theta \in [\delta, 2\pi - \delta]$ as $r \to R$.*

Proof. An elementary calculation shows that

$$P_{R,r}(\theta) = \operatorname{Re}\left\{\frac{R + r\,e^{i\theta}}{R - r\,e^{i\theta}}\right\} = \operatorname{Re}\left\{\frac{R + z}{R - z}\right\}.$$

Hence by (ii) of 6.1, $P_{R,r}$ is harmonic on $S(0, R)$.

The expansion

$$\frac{R + z}{R - z} = 1 + 2\frac{z}{R} + 2\left(\frac{z}{R}\right)^2 + \cdots = 1 + 2\sum_{k=1}^{\infty}\left(\frac{z}{R}\right)^k$$

is uniformly convergent for $|z| = r < R$, and shows that

$$P_{R,r}(\theta) = 1 + 2\sum_{k=1}^{\infty}\left(\frac{r}{R}\right)^k \cos k\theta.$$

In particular (a) follows on integrating term-by-term.

To prove (b), let $\theta \in [\delta, 2\pi - \delta]$, when

$$\begin{aligned} R^2 - 2Rr\cos\theta + r^2 &\geqslant R^2 - 2Rr\cos\delta + r^2 \\ &= (R\sin\delta)^2 + (R\cos\delta - r)^2 \geqslant (R\sin\delta)^2. \end{aligned}$$

Hence

$$0 < P_{R,r}(\theta) \leqslant \frac{R^2 - r^2}{(R\sin\delta)^2} \to 0 \quad \text{uniformly as} \quad r \to R.$$

Theorem 6.6. *Suppose that g is a continuous real valued function on $C(a, R)$. Define $P(g)$ on $S(a, R)$ by*

$$P(g)(a + r\,e^{i\theta}) = \frac{1}{2\pi}\int_0^{2\pi} P_{R,r}(\theta - \phi)g(a + R\,e^{i\phi})\,d\phi, \quad 0 \leqslant r < R.$$

Then $P(g)$ is harmonic on $S(a, R)$, and $P(g)(a + r\,e^{i\theta}) \to g(a + R\,e^{i\theta})$ uniformly as $r \to R$.

Proof. A calculation similar to that in 6.5 shows that

$$P_{R,r}(\theta - \phi) = \mathrm{Re}\left\{\frac{R\,e^{i\phi} + r\,e^{i\theta}}{R\,e^{i\phi} - r\,e^{i\theta}}\right\},$$

and hence by 6.1 part (ii), that $P_{R,r}(\theta - \phi)$ is harmonic as a function of $r\,e^{i\theta}$. In other words the mean value of $P_{R,r}(\theta - \phi)$ taken round any circle $C(b, s) \subset S(a, R)$ is equal to the value of $P_{R,r}(\theta - \phi)$ when $r\,e^{i\theta} = b$.

The corresponding property for $P(g)$ now follows on interchanging the order of the two integrations.

To show that $P(g)(a + r\,e^{i\theta}) \to g(a + R\,e^{i\theta})$ uniformly as $r \to R$, we make use of the fact that g, being continuous, is necessarily uniformly continuous for $0 \leqslant \theta \leqslant 2\pi$ (sec 1.4 (i)). This means that given $\varepsilon > 0$, there is a corresponding $\delta > 0$ such that (writing $g_1(\theta) = g(a + R\,e^{i\theta})$)

$$|g_1(\theta) - g_1(\theta')| < \varepsilon \quad \text{if} \quad |R\,e^{i\theta} - R\,e^{i\theta'}| < \delta.$$

If follows from (a) of 6.5 that

$$P(g)(a + r\,e^{i\theta}) - g(a + R\,e^{i\theta})$$

$$= \frac{1}{2\pi}\int_0^{2\pi} P_{R,r}(\theta - \phi)(g(a + R\,e^{i\phi}) - g(a + R\,e^{i\theta}))\,d\phi.$$

We divide the range of integration into two parts, I_1 and I_2 say, where I_1 contains those values of ϕ with $|R\,e^{i\theta} - R\,e^{i\phi}| < \delta$, and I_2 the rest.

Then the integral over I_1 is at most

$$\frac{\varepsilon}{2\pi}\int_{I_1} P_{R,r}(\theta - \phi)\,d\phi \leqslant \varepsilon,$$

while the integral over I_2 is at most

$$2(\sup|g|)\,\frac{1}{2\pi}\int_{I_2} P_{R,r}(\theta - \phi)\,d\phi$$

which tends to zero as $r \to R$ by (b) of 6.5.

Corollary 6.7. *Let u be harmonic on G, and $\bar{S}(a, R) \subset G$. Then if $0 \leqslant r < R$,*

$$u(a + r\,e^{i\theta}) = \frac{1}{2\pi}\int_0^{2\pi} P_{R,r}(\theta - \phi)u(a + R\,e^{i\phi})\,d\phi.$$

Proof. Let v be defined as equal to u on $C(a, R)$ and $P(u)$ on $S(a, R)$. Theorem 6.6 tells us that v is harmonic on $S(a, R)$ and continuous on $\bar{S}(a, R)$. It follows from (ii) of 6.3 that $u = v$ on $S(a, R)$, or in other words $u = P(u)$ as was to be proved.

Theorem 6.8. *Let u be harmonic on $S(a, R)$. Then there exists a regular function f on $S(a, R)$, uniquely determined to within a purely imaginary additive constant, with $u = \mathrm{Re}\,(f)$.*

In particular a harmonic function must haue partial derivatives of all orders, and be a solution of Laplace's equation.

Proof. We know that

$$P_{R,r}(\theta - \phi) = \mathrm{Re}\left\{\frac{R\,e^{i\phi} + r\,e^{i\theta}}{R\,e^{i\phi} - r\,e^{i\theta}}\right\};$$

given $r < R$, choose $r < R_1 < R$. It follows from 6.7 that since u is assumed real valued,

$$u(a + r\,e^{i\theta}) = \mathrm{Re}\left\{\frac{1}{2\pi}\int_0^{2\pi}\frac{R_1\,e^{i\phi} + r\,e^{i\theta}}{R_1\,e^{i\phi} - r\,e^{i\theta}}\,u(a + R_1\,e^{i\phi})\,d\phi\right\}$$
$$= \mathrm{Re}\,\{F_1\}, \quad \text{say.}$$

But F_1 is evidently regular in $r\,e^{i\theta}$, and u is its real part: suppose F_2 corresponded similarly to R_2, with $R_1 < R_2 < R$. Then

$$\mathrm{Re}\,\{F_1\} = \mathrm{Re}\,\{F_2\} = u \quad \text{on} \quad S(a, R_1),$$

and hence by the Cauchy–Riemann equations, $F_1 - F_2$ is an imaginary constant on $S(a, R_1)$. But the integral formulae for F_1 and F_2 show that $F_1(a) = F_2(a) = u(a)$, and hence $F_1 = F_2$ on $S(a, R_1)$. It follows that if we define $f = F_1$ for each choice of R_1 between r and R, then f will be well defined, and u its real part. The same argument shows that any other regular function f_1 with $u = \mathrm{Re}\,\{f_1\}$ can differ from f by at most an imaginary constant.

The final assertion is now a consequence of 6.1.

If we attempt to extend this result to domains other than a disc, we encounter the same kind of difficulty as was found in connection with Cauchy's theorem: some restriction on the shape of the domain is necessary. Theorem 6.9 below gives a sufficient condition.

Notice that we have solved a particular case of the Dirichlet problem, namely given a particular region G and a function g which is continuous on the boundary of G, find a continuous function u on $\bar{G}$ which is harmonic on G, and equal to g on ∂G. Theorem 6.6 shows that this can be done if G is a disc and we take $u = P(g)$. The problem for more general regions is much more difficult; an elementary treatment can be found in the book by Ahlfors mentioned in the bibliography.

§2. HARMONIC CONJUGATES

We begin by proving that if a function u is harmonic on any starred open set then the result corresponding to theorem 6.8 must be valid. We also show that a solution of Laplace's equation in such a set is necessarily harmonic.

Theorem 6.9. *Let G be an open set, starred with respect to a, and let u be harmonic on G.*

Then there is a unique harmonic function v satisfying $v(a) = 0$, for which $u + iv$ is regular on G.

Proof. Since u is harmonic, theorem 6.8 shows that it has continuous derivatives of all orders, and satisfies

$$\frac{\partial^2 u}{\partial x^2} + \frac{\partial^2 u}{\partial y^2} = 0.$$

In particular,

$$\frac{\partial u}{\partial x} - i\frac{\partial u}{\partial y} \; (= f, \text{ say})$$

is regular by 1.11.

Since G is starred, there is a unique primitive F for f with $F(a) = u(a)$ by the proof of 2.3. Hence $\operatorname{Re} F = u$, and $\operatorname{Im} F$ is the required harmonic conjugate v.

It is easily verified that for each $z \in G$, one explicit representation for v is given by

$$v(z) = (x - c) \int_0^1 \frac{\partial u}{\partial x} (a + t(z - a))\, dt$$
$$+ (y - d) \int_0^1 \frac{\partial u}{\partial y} (a + t(z - u))\, dt.$$

where $z = x + iy$, $a = c + id$, and we have written $(\partial u/\partial x)(w)$ in place of $(\partial u/\partial x)(\text{Re } w, \text{Im } w)$.

It follows from the results of section 1 that if u is harmonic on G, and $\bar{S}(a, R) \subset G$ (or even harmonic on $S(a, R)$ and continuous on $\bar{S}(a, R)$) then it follows that for $0 \leqslant r < R$,

$$u(a + r\, e^{i\theta}) = \frac{1}{2\pi} \int_0^{2\pi} P_{R,r}(\theta - \phi) u(a + R\, e^{i\phi})\, d\phi.$$

From this and the fact that

$$P_{R,r}(\theta - \phi) = \text{Re} \left\{ \frac{R\, e^{i\phi} + r\, e^{i\theta}}{R\, e^{i\phi} - r\, e^{i\theta}} \right\},$$

we deduce that another expression for v may be obtained by writing

$$Q_{R,r}(\theta - \phi) = \text{Im} \left\{ \frac{R\, e^{i\phi} + r\, e^{i\theta}}{R\, e^{i\phi} - r\, e^{i\theta}} \right\} = \frac{2Rr \sin(\theta - \phi)}{R^2 - 2Rr \cos(\theta - \phi) + r^2}$$

and hence

$$v(a + r\, e^{i\theta}) = \frac{1}{2\pi} \int_0^{2\pi} Q_{R,r}(\theta - \phi) u(a + R\, e^{i\phi})\, d\phi.$$

Corollary 6.10. *Let u be a function with continuous partial derivatives of the second order, and which satisfies Laplace's equation on an open set G. Then u is harmonic in G.*

Proof. Since

$$\frac{\partial^2 u}{\partial x^2} + \frac{\partial^2 u}{\partial y^2} = 0,$$

it follows as above that

$$f = \frac{\partial u}{\partial x} - i \frac{\partial u}{\partial y}$$

is regular in G. If $\bar{S}(a, R) \subset G$, then in the same way as in 6.9, we have a function F regular on $S(a, R)$ with $F' = f$, and so $\operatorname{Re} F = u$. In particular it now follows from 6.1 that

$$u(a) = \frac{1}{2\pi}\int_0^{2\pi} u(a + R\,e^{i\theta})\,d\theta, \quad \text{so } u \text{ is harmonic in } G.$$

Our final result shows that the mapping $u \to v$ which takes a harmonic function to its conjugate is continuous relative to a pth mean value of f which we now define.

Definition 6.11. Let f be continuous on $C(a, r)$. Define

$$M_p(f, r) = \left(\frac{1}{2\pi}\int_0^{2\pi} |f(a + r\,e^{i\theta})|^p\,d\theta\right)^{1/p} \quad \text{for} \quad 0 < p < \infty.$$

For our purposes r will generally be fixed, and we shall simply write $M_p(f)$ for $M_p(f, r)$.

In the case $p \geqslant 1$, $M_p(f)$ is known as the pth mean, or p-norm of f. (Readers who know some linear space theory will not need to be reminded of the norm-properties of M_p—however, we shall not make use of these here.)

Our aim is to prove the following.

Theorem 6.12. (Riesz conjugation theorem). *Let* $1 < p < \infty$, *and let*

$$A_p = \max\left[\tan\left(\frac{\pi}{2p}\right), \cot\left(\frac{\pi}{2p}\right)\right].$$

Then if u is a positive harmonic function on $S(a, R)$, and v is the harmonic conjugate of u on $S(a, R)$ with $v(a) = 0$, then for each $r < R$, we have

$$M_p(v, r) \leqslant A_p M_p(u, r).$$

The proof of this result will come after the proof of a couple of elementary results on trigonometric functions. The result is in fact valid without the restriction that u is positive—for this the reader should consult the paper by S. K. Pichorides in *Studia Math.* (1972), Vol. XLIV, No. 2. It can be shown by means of examples that the constant A_p is the best possible.

We shall restrict our attention to the case $1 < p \leqslant 2$, when $A_p = \tan(\pi/(2p))$: an analogous argument proves the result in the case $p \geqslant 2$.

Lemma 6.13. (i) *Let* $1 < p \leqslant 2$. *Then*

$$f(x) = \frac{\sin px}{p \sin x(\cos x)^{p-1}}$$

is a non-decreasing function of x *on* $[0, \frac{1}{2}\pi)$.

(ii) *With the same values of* p, *write* $b = \pi/(2p)$, *so that* $\pi/4 \leqslant b < \pi/2$. *Then for* $\theta \in [-\frac{1}{2}\pi, \frac{1}{2}\pi]$,

$$(\sin b)^{p-1} \sec b \cos p\theta \leqslant (\tan b)^p \cos^p \theta - (|\sin \theta|)^p.$$

Proof. (i) Let y be a positive real number. Then two integrations by parts show that

$$(1 + iy)^p = 1 + ipy - p(p-1) \int_0^y (y - t)(1 + it)^{p-2}\, dt.$$

[The observant reader will have noticed that we are using non-integer powers of complex numbers which we have so far carefully avoided. A quick look at Appendix C should reassure him that the familiar properties remain true—if correctly interpreted!]

If we now put $y = \tan x$, $t = \tan u$, then we obtain

$$\frac{\cos px + i \sin px}{\cos^p x} = 1 + ip \tan x$$

$$- p(p-1) \int_0^x (\tan x - \tan u) \frac{\cos(p-2)u + i \sin(p-2)u}{\cos^p u}\, du$$

for $0 \leqslant x < \frac{1}{2}\pi$.

Equating imaginary parts now shows that

$$\frac{\sin px}{\cos^p x}$$

$$= p \tan x + p(p-1) \int_0^x (\tan x - \tan u) \sin(2-p)u\, (\sec u)^p\, du.$$

or

$$\frac{\sin px}{p \sin x\, (\cos x)^{p-1}}$$

$$= 1 + (p-1) \int_0^x (1 - \tan u \cot x) \sin(2-p)u\, (\sec u)^p\, du.$$

The right-hand side is now obviously a non-decreasing function of x.

(ii) Since both sides of the required inequality are even, it is sufficient to consider $\theta \in [0, \frac{1}{2}\pi]$. Writing $b = \pi/(2p)$, and noticing that $\sin pb = 1$, it is easy to verify that the equation

$$\tan^p b + \tan^{p-2} \theta = \frac{(\sin b)^{p-1}}{\cos b} \cdot \frac{\sin p\theta}{\sin \theta (\cos \theta)^{p-1}}$$

is satisfied when $\theta = b$.

However, for $\theta \in [0, \frac{1}{2}\pi)$, the left-hand side is a non-increasing function of θ ($p \leqslant 2$), while the right-hand side is non-decreasing by (i) above.

Hence if $\theta \in [b, \frac{1}{2}\pi)$,

$$\tan^p b + \tan^{p-2} \theta \geqslant \frac{(\sin b)^{p-1}}{\cos b} \cdot \frac{\sin p\theta}{\sin \theta (\cos \theta)^{p-1}},$$

while the inequality is reversed if $\theta \in [0, b]$.

This may be rewritten

$$\tan^p b \sin \theta (\cos \theta)^{p-1} + (\sin \theta)^{p-1} \cos \theta \leqslant \frac{(\sin b)^{p-1}}{\cos b} \sin p\theta,$$

and the result follows on integrating from b to θ (in case $\theta \in [b, \frac{1}{2}\pi]$), or θ to b (if $\theta \in [0, b]$).

Proof of theorem 6.12. Let u be positive and harmonic in $S(a, R)$, v be its harmonic conjugate with $v(a) = 0$, and suppose $r < R$ is given.

For $0 \leqslant \alpha \leqslant 2\pi$, write $u(r\, e^{i\alpha}) + iv(r\, e^{i\alpha}) = \rho\, e^{i\theta}$. If we assume u does not vanish identically, then we must have $u > 0$ on $S(a, R)$ (otherwise the mean value property would be violated). Hence $\rho > 0$, and we may take $|\theta| < \frac{1}{2}\pi$. We can also assume $a = 0$ without loss of generality.

Then

$$\begin{aligned}(M_p(v))^p &= \frac{1}{2\pi} \int_0^{2\pi} |v(r\, e^{i\alpha})|^p \, d\alpha \\ &= \frac{1}{2\pi} \int_0^{2\pi} \rho^p |\sin \theta|^p \, d\alpha \\ &\leqslant \frac{1}{2\pi} \int_0^{2\pi} \rho^p \{\tan^p b \cos^p \theta - (\sin b)^{p-1} \sec b \cos p\theta\} \, d\alpha\end{aligned}$$

on applying (ii) of 6.13.

The first of these terms is

$$\tan^p b \frac{1}{2\pi} \int_0^{2\pi} \rho^p \cos^p \theta \, d\alpha = \tan^p b \, (M_p(u))^p.$$

The second term is a constant multiple of

$$\frac{1}{2\pi} \int_0^{2\pi} \rho^p \cos p\theta \, d\alpha = \operatorname{Re} \left\{ \frac{1}{2\pi} \int_0^{2\pi} (u + iv)^p \, d\alpha \right\}$$

$$= (u(0))^p > 0.$$

Hence it follows that if u is positive and does not vanish identically,

$M_p(v) < \tan b \, M_p(u)$, as was to be proved.

EXERCISES FOR CHAPTER 6

1. Show that if u is a homogeneous polynomial in x, y of degree two (that is $u(x, y) = ax^2 + bxy + cy^2$) which is harmonic, then u must be a linear combination of $u_1 = x^2 - y^2$, and $u_2 = xy$.

Find the corresponding polynomials of degrees three and four.

2. Prove that a uniform limit of harmonic functions is harmonic.

Define g on $C(0, 1)$ by $g(z) = \pm 1$ according as $\operatorname{Re} z$ is >0 or <0, and $g(\pm i) = 0$.

Can you find a sequence of functions which are harmonic on $\bar{S}(0, 1)$ and converge (non-uniformly!) to g on $C(0, 1)$?
Hint: consider $P(g)$.

3. For $0 \leqslant r < 1$, and $0 \leqslant \theta < 2\pi$, find:

(a) $\sup\limits_{0 \leqslant r \leqslant 1} P_r(\theta)$ as a function of θ, and

(b) $\sup\limits_{0 \leqslant \theta \leqslant 2\pi} P_r(\theta)$ as a function of r.

Show that the contours '$P_r(\theta) =$ constant' are circles which are all mutually tangent at the point $1 \in C(0, 1)$, and lie in $\bar{S}(0, 1)$.

Investigate $Q_r(\theta) = 2r \sin \theta(1 - 2r \cos \theta + r^2)^{-1}$ similarly.

4. Show that if $r = |z|$, the function $\log r$ is harmonic on $\mathbf{C} \setminus \{0\}$, r^α is harmonic if and only if $\alpha = 0$, and xr^α ($x = \operatorname{Re} z$) is harmonic if and only if $\alpha = 0$ or -2.

5. For each of the following functions, verify that it is harmonic and find a harmonic conjugate subject to the conditions stated.

(i) $u(x, y) = \sinh x \cos y$, $z = x + iy \in \mathbf{C}$, $v(0) = 0$,
(ii) $u(x, y) = xy(x^2 + y^2)^{-2}$, $z = x + iy \in \mathbf{C} \setminus \{0\}$, $v(1) = 0$,
(iii) $u(x, y) = e^y \sin x$, $z = x + iy \in \mathbf{C}$, $v(0) = 0$.

6. Prove that, if u is harmonic on an open set G, and $S(a, R) \subset G$, then there exist uniquely determined coefficients $(a_n)_{n=0}^{\infty}$, $(b_n)_{n=1}^{\infty}$, with

$$u(z) = a_0 + \sum_{n=1}^{\infty} (a_n \cos n\theta + b_n \sin n\theta) r^n,$$

where

$$z = a + r\, e^{i\theta} \in S(a, R),$$

and the series is uniformly convergent with respect to $\theta \in [0, 2\pi]$, for each $r < R$.

Find formulae for a_n, b_n in terms of an integral of u around $C(a, r)$.

Appendix A
The Regulated Integral

This appendix gives a brief exposition of the so-called regulated integral, which first came to prominence in the book *Foundations of Modern Analysis*, by J. Dieudonné, mentioned in the bibliography. It has the great advantage over the traditional Riemann theory of describing from the outset a readily identifiable class of functions (the regulated functions) which are to be integrated. This class includes all continuous functions and all monotone functions. The definitions are slightly modified from those given by Dieudonné in order to facilitate a subsequent generalization to Lebesgue integration.

Definition A.1. (The class of regulated functions.) Let f be a real or complex valued function on the interval $[a, b]$. We use the notation

$$f(x+) \quad \text{for} \quad \lim_{h \to 0+} f(x+h), \quad \text{and} \quad f(x-) \quad \text{for} \quad \lim_{h \to 0+} f(x-h):$$

these are the right- and left-handed limits of f at x, respectively. Say f is regulated if $f(x+)$ exists for all $x \in [a, b)$, and $f(x-)$ exists for all $x \in (a, b]$. These limits need not in general be equal either to each other, or to $f(x)$.

Define

$$j(x) = \max\left(|f(x+) - f(x)|, |f(x-) - f(x)|, |f(x+) - f(x-)|\right).$$

Obviously f is continuous at x if and only if $j(x) = 0$.

Lemma A.2.

(i) *Any continuous, or monotone function is regulated.*

(ii) *If f, g are regulated, so are $f + g$, fg, $f \vee g$, and $f \wedge g$.*

(iii) *A regulated function is bounded on $[a, b]$.*

(iv) *For any $h > 0$, the number of points where $j(x) > h$ is finite. Hence a regulated function is continuous except on a set which is at most denumerably infinite.*

Proof. (i) A continuous function has right- and left-hand limits equal to $f(x)$: a monotone function has right- and left-hand limits which satisfy $f(x-) \leqslant f(x) \leqslant f(x+)$. Both are regulated functions.

(ii) Is immediate from the corresponding properties of limits.

(iii) If $x \in [a, b]$, then existence of right- and left-hand limits at x shows that f is bounded on a set of the form $(x - \delta, x + \delta) \cap [a, b]$ for some $\delta > 0$. The Heine–Borel property (1.2) now shows that f is bounded on $[a, b]$.

(iv) Suppose that for some $h > 0$ there were infinitely many values of x with $j(x) > h$. The Bolzano–Weierstrass property (1.2) shows that there is a point $x_0 \in [a, b]$ and a sequence (which may be assumed monotone) of points (x_n) which converge to x_0, with $j(x_n) > h > 0$ for each n. If the sequence increases to x_0, then f cannot have a left-hand limit at x_0 and so is not regulated, and similarly if the sequence decreases to x_0.

Hence the set $\{x : j(x) > h > 0\}$ must be finite. The set of discontinuities of f is simply

$$\bigcup_{n=1}^{\infty} \{x : j(x) > 1/n\},$$

and so is at most denumerably infinite.

Definition A.3. A step function s on $[a, b]$ is a real valued function which is constant on open intervals of the form (t_{i-1}, t_i), $i = 1, 2, \ldots, n$, where

$$a = t_0 < t_1 < \cdots < t_{i-1} < t_i < \cdots < t_n = b.$$

The value of s at t_i $(i = 0, 1, \ldots, n)$ may be any real number.

If $s(x) = \alpha_i$ for $x \in (t_{i-1}, t_i)$, we shall write

$$s = \sum_{i=1}^{n} \alpha_i \chi(t_{i-1}, t_i)$$

and neglect the value of s at t_i $(i = 0, 1, \ldots, n)$. In order to make this representation unique we assume that s is actually discontinuous at each $t_i \neq a, b$: that is either

$$\alpha_i \neq \alpha_{i+1} \quad \text{or} \quad \alpha_i = \alpha_{i+1} \neq s(t_i).$$

If s is constant on (a, b), then $t_0 = a$, $t_1 = b$.

Evidently a step function is regulated.

Lemma A.4. *Let f be a positive regulated function on $[a, b]$. Then for any $\varepsilon > 0$, there is a positive step function s on $[a, b]$, such that $f(x) \geqslant s(x) \geqslant f(x) - \varepsilon$ for all $x \in [a, b]$.*

Proof. Let $\varepsilon > 0$ be given.

Let E be the set of points c in $[a, b]$ for which there exists a function s with the required property on $[a, c]$. E is an interval (which may consist of a alone!) having a for its left-hand-end point: let $c_0 = \sup E$, and suppose $a \leqslant c_0 < b$.

Since f is regulated, $f(c_0-)$ and $f(c_0+)$ exist (or only the latter if $c_0 = a$). Then there is a $\delta > 0$ such that

$$|f(x) - f(c_0-)| < \tfrac{1}{2}\varepsilon \quad \text{if} \quad x \in (c_0 - \delta, c_0)$$

and

$$|f(x) - f(c_0+)| < \tfrac{1}{2}\varepsilon \quad \text{if} \quad x \in (c_0, c_0 + \delta).$$

Hence if we define s_0 by

$$s_0(x) = (f(c_0-) - \tfrac{1}{2}\varepsilon) \vee 0 \quad \text{on} \quad (c_0 - \delta, c_0), \quad s_0(c_0) = f(c_0)$$

and

$$s_0(x) = (f(c_0+) - \tfrac{1}{2}\varepsilon) \vee 0 \quad \text{if} \quad x \in (c_0, c_0 + \delta),$$

then s_0 is a step function with the required property on $(c_0 - \delta, c_0 + \delta)$ (or $[a, a + \delta)$ if $c_0 = a$). The definition of c_0 shows that there is a step function s_1 with the required property on $[a, c_0 - \delta]$ (this also is not needed in the case $c_0 = a$). Combining s_1 on $[a, c_0 - \delta]$ with s_0 on $(c_0 - \delta, c_0 + \delta)$ gives a step function with the required property on $[a, c_0 + \delta)$ and so contradicts the definition of c_0.

It follows that c_0 must equal b, and the above construction gives a step function with the required property on $[a, b - \delta]$ and on $(b - \delta, b]$, and so on the whole of $[a, b]$.

Definition A.5. Let s be a positive step function on $[a, b]$. As in A.3, we write

$$s = \sum_{i=1}^{n} \alpha_i \chi(t_{i-1}, t_i), \qquad \alpha_i \geqslant 0, \qquad i = 1, 2, \ldots, n.$$

We define the integral of s over $[a, b]$ by

$$\int_a^b s = \sum_{i=1}^{n} \alpha_i (t_i - t_{i-1}).$$

This real number is uniquely determined by s, as noted in A.3 and is independent of the values of s at the discontinuities t_i, $i = 1, 2, \ldots, n - 1$.

Lemma A.6. *Let s_1, s_2 be positive step functions, with $s_1(x) \leqslant s_2(x)$ for all $x \in [a, b]$.*

Then $\int_a^b s_1 \leqslant \int_a^b s_2$.

Proof. Suppose that

$$s_1 = \sum_{i=1}^{n} \alpha_i \chi(t_{i-1}, t_i), \quad \text{and} \quad s_2 = \sum_{j=1}^{m} \beta_j \chi(s_{j-1}, s_j).$$

For $i = 1, 2, \ldots, n$, $j = 1, 2, \ldots, m$, let E_{ij} denote the interval

$$(t_{i-1}, t_i) \cap (s_{j-1}, s_j)$$

(which may well be empty), and l_{ij} be the length of E_{ij} ($l_{ij} = 0$ if $E_{ij} = \varnothing$). Then

$$t_i - t_{i-1} = \sum_{j=1}^{m} l_{ij}, \qquad s_j - s_{j-1} = \sum_{i=1}^{n} l_{ij},$$

and since $s_1 \leqslant s_2$, $\alpha_i \leqslant \beta_j$ if $l_{ij} \neq 0$.

It follows that

$$\int_a^b s_1 = \sum_{i=1}^{n} \alpha_i (t_. - t_{i-1}) = \sum_{i=1}^{n} \alpha_i \sum_{j=1}^{m} l_{ij} = \sum_{j=1}^{m} \sum_{i=1}^{n} \alpha_i l_{ij}$$

$$\leqslant \sum_{j=1}^{m} \sum_{i=1}^{n} \beta_j l_{ij} = \sum_{j=1}^{m} \beta_j (s_j - s_{j-1}) = \int_a^b s_2, \quad \text{as required.}$$

A similar argument proves

Lemma A.7. *Let s_1 and s_2 be positive step functions on $[a, b]$.*

Then $\int_a^b (s_1 + s_2) = \int_a^b s_1 + \int_a^b s_2$.

Proof. Using the notation introduced in A.6, we have

$$\int_a^b (s_1 + s_2) = \sum_{i=1}^{n} \sum_{j=1}^{m} (\alpha_i + \beta_j) l_{ij} = \sum_{i=1}^{n} \alpha_i \sum_{j=1}^{m} l_{ij} + \sum_{j=1}^{m} \beta_j \sum_{i=1}^{n} l_{ij}$$
$$= \sum_{i=1}^{n} \alpha_i (t_i - t_{i-1}) + \sum_{j=1}^{m} \beta_j (s_j - s_{j-1}) = \int_a^b s_1 + \int_a^b s_2.$$

Definition A.8. Let f be a positive regulated function on $[a, b]$. By lemma A.2 (iii) f is bounded above, say $f(x) \leqslant M$ for all $x \in [a, b]$. It follows that if s is a step function with $0 \leqslant s \leqslant f$, then $s(x) \leqslant M$ on $[a, b]$ and hence that (by A.6) $\int^b s$ is bounded above by $M(b - a)$.

This allows us to define the integral of f over $[a, b]$ by

$$\int_a^b f = \sup \left\{ \int_a^b s : s \quad \text{is a step function with} \quad 0 \leqslant s \leqslant f \right\}.$$

Notice that definition A.5 says that the integral of a positive step function is the sum of the areas of the finite number of rectangles which are enclosed between it and the axis. Our new definition for a general positive regulated function simply says that the integral is the supremum of all such sums of rectangles enclosed between the function and the axis. As such it embodies the natural idea of the integral as the 'area under the curve'.

For a real valued regulated function which is not necessarily positive, let $f^+ = f \vee 0$, and $f^- = -(f \wedge 0)$, and define

$$\int_a^b f = \int_a^b f^+ - \int_a^b f^-.$$

If f is complex valued and regulated, then $f = f_1 + if_2$ where (as may be easily verified) f_1 and f_2 are real valued and regulated. In this case we define

$$\int_a^b f = \int_a^b f_1 + i \int_a^b f_2.$$

If $a > b$, define $\int_a^b f = - \int_b^a f$.

There is a consistency problem which arises from this definition when f is a positive step function, since we then have two definitions (A.5 and A.8) for $\int_a^b f$. Suppose that in accordance with A.8, s is a positive step function, $s \leqslant f$; by A.5 we have that $\int_a^b s \leqslant \int_a^b f$.

But if f is a step function, we may take $s = f$ in A.8 to show that the supremum is attained and hence that the two definitions give the same value to $\int_a^b f$.

Theorem A.9. *For a regulated function on an interval in* $\mathbf{R}$, $\int_a^b f$ *is monotone and linear as a function of* f, *and additive as a function of* $[a, b]$.

More precisely, we have

(i) *if* f_1 *and* f_2 *are real and regulated, and* $f_1 \leqslant f_2$, *then*

$$\int_a^b f_1 \leqslant \int_a^b f_2,$$

(ii) *if* f_1 *and* f_2 *are regulated,* $\int_a^b (f_1 + f_2) = \int_a^b f_1 + \int_a^b f_2$,

(iii) *if* c *is real or complex and* f *is regulated, then* $\int_a^b cf = c \int_a^b f$,

(iv) *if* $a < c < b$, *and* f *is regulated, then* $\int_a^b f = \int_a^c f + \int_c^b f$.

Proof. (i) If $f_1 \leqslant f_2$, then $0 \leqslant f_1^+ \leqslant f_2^+$, and $0 \leqslant f_2^- \leqslant f_1^-$. It follows immediately from definition A.8 that

$$0 \leqslant \int_a^b f_1^+ \leqslant \int_a^b f_2^+ \quad \text{and} \quad 0 \leqslant \int_a^b f_2^- \leqslant \int_a^b f_1^-.$$

Hence

$$\int_a^b f_1^+ - \int_a^b f_1^- \leqslant \int_a^b f_2^+ - \int_a^b f_2^- \quad \text{as required.}$$

(ii) Suppose firstly that f_1 and f_2 are positive. Let s_1 and s_2 be any step functions with $0 \leqslant s_i \leqslant f_i$, $i = 1, 2$.

Then

$$\int_a^b s_1 + \int_a^b s_2 = \int_a^b (s_1 + s_2) \leqslant \int_a^b (f_1 + f_2),$$

using A.7 and A.8. It follows on taking suprema over s_1 and s_2 that

$$\int_a^b f_1 + \int_a^b f_2 \leqslant \int_a^b (f_1 + f_2).$$

To prove the reverse inequality we make use of A.4.

Suppose that $\varepsilon > 0$ is given, and that s_1, s_2 are positive simple functions with

$$s_i \leqslant f_i \leqslant s_i + \varepsilon \quad \text{for} \quad i = 1, 2.$$

It follows that $f_1 + f_2 \leqslant s_1 + s_2 + 2\varepsilon$, and so from (i) above that

$$\begin{aligned}\int_a^b (f_1 + f_2) \leqslant \int_a^b (s_1 + s_2 + 2\varepsilon) &= \int_a^b s_1 + \int_a^b s_2 + \int_a^b 2\varepsilon \\ &= \int_a^b s_1 + \int_a^b s_2 + 2\varepsilon(b - a) \\ &\leqslant \int_a^b f_1 + \int_a^b f_2 + 2\varepsilon(b - a),\end{aligned}$$

where we have used in turn A.7, A.5 and A.8. The reverse inequality now follows from the fact that ε is arbitrary. We now show that if f is a regulated function, and $f = g - h$, where g and h are positive and regulated, then

$$\int_a^b f = \int_a^b g - \int_a^b h.$$

For $g - h = f = f^+ - f^-$, and so $g + f^- = h + f^+$. It follows from what we have just proved for positive functions that

$$\int_a^b f + \int_a^b f^- = \int_a^b h + \int_a^b f^+.$$

Hence

$$\int_a^b f = \int_a^b f^+ - \int_a^b f^- = \int_a^b h - \int_a^b g.$$

The case when f_1, f_2 are assumed real, but not necessarily positive now follows on taking

$$h = f_1^+ + f_2^+ \quad \text{and} \quad g = f_1^- + f_2^-.$$

For $h - g = f_1 + f_2$, and so,

$$\begin{aligned}\int_a^b (f_1 + f_2) = \int_a^b h - \int_a^b g &= \left(\int_a^b f_1^+ + \int_a^b f_1^+\right) - \left(\int_a^b f_1^- + \int_a^b f_2^-\right) \\ &= \left(\int_a^b f_1^+ - \int_a^b f_1^-\right) + \left(\int_a^b f_2^+ - \int_a^b f_2^-\right) \\ &= \int_a^b f_1 + \int_a^b f_2,\end{aligned}$$

which gives the result in this case. The general case where f_1 and f_2 may take complex values follows at once on considering their real and imaginary parts.

(iii) If $c \geqslant 0$, and s is a positive step function, then it is immediate from A.5 that

$$c \int_a^b s = \int_a^b cs,$$

and the corresponding result for positive regulated f follows on taking suprema. If f is real valued and $c \geqslant 0$,

$$\int_a^b cf = \int_a^b (cf)^+ - \int_a^b (cf)^- = c \int_a^b f^+ - c \int_a^b f^- = c \int_a^b f.$$

If $c < 0$ and f is real valued,

$$(cf)^+ = |c| f^- \quad \text{and} \quad (cf)^- = |c| f^+,$$

so that

$$\int_a^b cf = |c| \left(\int_a^b f^- - \int_a^b f^+ \right) = |c| \left(- \int_a^b f \right) = c \int_a^b f.$$

The general case when c and f may be complex follows on considering their real and imaginary parts.

(iv) Let f be positive, and s, s' be step functions such that $0 \leqslant s \leqslant f$ on $[a, c]$, $0 \leqslant s' \leqslant f$ on $[c, b]$.

Then if $s'' = s$ on $[a, c]$ and s' on $[c, b]$, it follows that s'' is a step function with $0 \leqslant s'' \leqslant f$ on $[a, b]$. Then

$$\int_a^b f \geqslant \int_a^b s'' = \int_a^c s + \int_c^b s',$$

on applying A.5 to s and s'. Hence on taking suprema,

$$\int_a^b f \geqslant \int_a^c f + \int_c^b f.$$

Conversely, suppose $\varepsilon > 0$ is given, and we take a positive step function t on $[a, b]$ with

$$\int_a^b f \geqslant \int_a^b t \geqslant \int_a^b f - \varepsilon.$$

Then if t_1, t_2 are the restrictions of t to $[a, c]$ and $[c, b]$ respectively, we have

$$\int_a^b f \leqslant \int_a^b t + \varepsilon = \int_a^c t_1 + \int_c^b t_2 + \varepsilon \leqslant \int_a^c f + \int_c^b f + \varepsilon,$$

and the result follows, since ε is arbitrary.

When f is not restricted to be positive the result follows at once on considering the positive and negative parts of $\text{Re}\,(f)$ and $\text{Im}\,(f)$.

We have now defined our integral on the class of regulated functions, and obtained its formal properties. We next deduce the fundamental theorems of calculus which establish the link with differentiation, and allow a number of useful ways of actually evaluating integrals.

Theorem A.10. (Fundamental theorems of calculus). (i) *Let f be regulated on $[a, b]$. Then $F(x) = \int_a^x f$ is continuous on $[a, b]$ and has a right derivative equal to $f(x+)$ at each point of $[a, b)$ and a left derivative equal to $f(x-)$ at each point of $(a, b]$. In particular, at every point of continuity of f in $[a, b]$, F is differentiable and $F'(x) = f(x)$.*

(ii) *Let g be a differentiable function on $[a, b]$ with g' continuous on* $[a, b]$. (The intermediate value property for derivatives shows that any regulated derivative must be continuous.) *Then*

$$\int_a^b g' = g(b) - g(a).$$

Proof. (i) We prove the statement about right derivatives: the statement about left derivatives is similar, and if these are equal, F is necessarily differentiable. We can suppose without loss of generality that the functions involved are real valued.

Suppose then that $x \in [a, b)$ and let $m = f(x+)$. Then for $\varepsilon > 0$, there is a $\delta > 0$ such that

$$m - \varepsilon < f(x + h) < m + \varepsilon \quad \text{if} \quad 0 < h < \delta.$$

It follows that

$$F(x + h) - F(x) = \int_a^{x+h} f - \int_a^x f = \int_x^{x+h} f$$

lies between

$$\int_x^{x+h} (m - \varepsilon) \quad \text{and} \quad \int_x^{x+h} (m + \varepsilon) \quad \text{if} \quad 0 < h < \delta.$$

In other words

$$(m - \varepsilon)h < F(x + h) - F(x) < (m + \varepsilon)h \quad \text{if} \quad 0 < h < \delta.$$

Hence

$$\left|\frac{1}{h}(F(x + h) - F(x)) - m\right| < \varepsilon \quad \text{if} \quad 0 < h < \delta,$$

and the right-hand derivatives of F at x exists and is equal to m.

(ii) Let g' be continuous on $[a, b]$, and $h(x) = \int_a^x g'$.

Then from (i) we know that $h'(x) = g'(x)$ for all $x \in [a, b]$.

It follows from the mean value theorem for derivatives that $h(x) - g(x)$ is constant on $[a, b]$, and since $h(a) = 0$,

$$h(x) = g(x) - g(a).$$

In particular

$$\int_a^b g' = h(b) = g(b) - g(a).$$

Definition A.11. Let f be continuous on $[a, b]$. A function F for which $F'(x) = f(x)$ for all $x \in [a, b]$ (including one-sided derivatives at the end points) is called a primitive function, or indefinite integral for f.

Part (i) of A.10 shows that $\int_a^x f$ is a primitive, and the proof of part (ii) shows that any two primitives can differ at most by an additive constant.

Theorem A.12. (Integration by parts and substitution). (i) *Let f, g be continuous on $[a, b]$ and F, G be primitive functions for f, g.*

Then

$$\int_a^b Fg = F(b)G(b) - F(a)G(a) - \int_a^b fG.$$

(ii) *Let g have a continuous derivative on $[a, b]$ and f be continuous on $g([a, b]) = \{y : y = g(x)$ for some $x \in [a, b]\}$.*

Then

$$\int_{g(a)}^{g(b)} f = \int_a^b (f \circ g)g'.$$

Proof. (i) The rule for differentiating a product of differentiable functions shows that $(FG)' = Fg + fG$.

Hence

$$\int_a^b Fg + \int_a^b fG = \int_a^b (FG)' = F(b)G(b) - F(a)G(a) \quad \text{as required.}$$

(ii) Let $F(x) = \int_{g(a)}^x f$, and notice that F is differentiable on $g([a, b])$, with $F'(x) = f(x)$.

Then $(F \circ g)' = (F' \circ g)g'$ by the chain rule for differentiation of composite functions, and so

$$\int_{g(a)}^{g(b)} f = F(g(b)) - F(g(a)) = \int_a^b (F \circ g)' = \int_a^b (F' \circ g)g' = \int_a^b (f \circ g)g'$$

as required.

Theorem A.13. *Let f be complex valued and regulated on $[a, b]$.*

Then

$$\left|\int_a^b f\right| \leqslant \int_a^b |f|.$$

Proof. If $\int_a^b f = 0$, the result is trivial: if not, choose c so that $|c| = 1$, and $c \int_a^b f = |\int_a^b f|$ is real and >0.

Then

$$\left|\int_a^b f\right| = c\int_a^b f = \int_a^b cf = \int_a^b \operatorname{Re}(cf) \leqslant \int_a^b |cf| = \int_a^b |f|.$$

Finally we have a result on limits which is used in Chapter 3.

Theorem A.14. *Let (f_n) be a sequence of regulated functions on $[a, b]$, and suppose $f_n \to f$ uniformly on $[a, b]$ as $n \to \infty$.*

Then f is also regulated, and

$$\int_a^b f_n \to \int_a^b f \quad \text{as} \quad n \to \infty.$$

Proof. To show that f is regulated we take a point $x \in [a, b)$, and show that f has a right-hand limit at x. A similar argument establishes that a left-hand limit exists for all $x \in (a, b]$, and hence that f is regulated.

Suppose then that $x \in [a, b)$, and that $\varepsilon > 0$ is given.

Then there is an n_0, depending on ε only, such that

$$|f_n(y) - f(y)| < \tfrac{1}{3}\varepsilon \quad \text{for all} \quad n \geqslant n_0, \quad \text{and all} \quad y \in [a, b].$$

In particular

$$|f_{n_0}(y) - f(y)| < \tfrac{1}{3}\varepsilon \quad \text{for all} \quad y \in [x, b].$$

Now f_{n_0} has a right-hand limit at x, and so there is a $\delta > 0$ depending on ε and n_0 (and so ultimately on ε alone) for which

$$|f_{n_0}(y) - f_{n_0}(y')| < \tfrac{1}{3}\varepsilon$$

if both y and y' are in $(x, x + \delta)$.

It follows that

$$\begin{aligned} |f(y) - f(y')| &\leqslant |f(y) - f_{n_0}(y)| + |f_{n_0}(y) - f_{n_0}(y')| + |f_{n_0}(y') - f(y')| \\ &< \tfrac{1}{3}\varepsilon \qquad\qquad\quad + \tfrac{1}{3}\varepsilon \qquad\qquad\qquad + \tfrac{1}{3}\varepsilon = \varepsilon. \end{aligned}$$

if both y and y' are in $(x, x + \delta)$.

The general principle of convergence (Cauchy criterion) now shows that f has a limit as y approaches x from the right.

The proof that the integrals converge is now almost immediate.

Suppose that $\varepsilon > 0$ is given, and $|f_n(x) - f(x)| < \varepsilon$ for all $n \geqslant n_0$, and $x \in [a, b]$. Then it follows from A.9 and A.13 that

$$\begin{aligned} \left| \int_a^b f_n - \int_a^b f \right| = \left| \int_a^b (f_n - f) \right| &\leqslant \int_a^b |f_n - f| \\ &\leqslant \varepsilon(b - a), \quad \text{for} \quad n \geqslant n_0, \end{aligned}$$

and the result is proved.

Appendix B
Some Topological Considerations

A reader who already has a fair knowledge of complex analysis will realize on reading the main text, and in particular Chapters 2 and 5, that although the theorems which we have proved are adequate for our purposes, they stop some way short of telling the whole story. In particular the restriction in Cauchy's theorem and the residue theorem to starred open sets is an artificiality which a more advanced treatment would remove. One of the purposes of this appendix is to introduce the more sophisticated notion of simple connectedness, which provides the proper setting for Cauchy's theorem: the other is to discuss the famous theorem of Jordan on closed non-self-intersecting curves.

§1. SIMPLE CONNECTEDNESS

Our aim in this section will be to introduce the condition (which will be both necessary and sufficient) on an open set G in $\mathbf{C}$ which will ensure that for every regular function f on G, and every closed path γ, with $\gamma^* \subset G$, we will have

$$\int_\gamma f(z)\,dz = 0.$$

It was pointed out in Chapter 2 that some sort of condition on the shape of G is necessary, since lemma 1.22 gives an example of an open set ($\mathbf{C} \setminus \{0\}$), a function (given by $f(z) = z^{-1}$) and a closed path $\gamma = C(0, 1)$ for which

$$\int_\gamma f(z)\,dz = 2\pi i.$$

If we examine this example, it seems that what is wrong is that the open set has a point missing (a 'hole' at the origin) and that the path γ encircles this point (or, more accurately, $n(\gamma, 0) \neq 0$). If we wish to avoid this situation we can proceed in one of two ways.

We have to suppose from now on that our open sets G are connected, and remind the reader of the properties of connectedness which are discussed in 1.17 and example 16 of Chapter 1. The first condition which might embody the idea that an open set should have no holes is that the complement of G (that is $\mathbf{C} \setminus G$) should be connected: for instance the annulus $\{z: 1 < |z| < 2\}$ has a conspicuous hole, and its complement is evidently not connected. However, if $G = \mathbf{C} \setminus \{0\}$, a set which we wish to exclude, $\mathbf{C} \setminus G = \{0\}$, which is connected. The answer to this difficulty is to take complements with respect to the extended complex plane $\mathbf{C}^*$, introduced in Chapter 4. In this case, if $G = \mathbf{C} \setminus \{0\}$, $\mathbf{C}^* \setminus G = \{0, \infty\}$, which is not connected. We make this our definition of 'holelessness':

Definition B.1. Let G be a connected open subset of $\mathbf{C}$. We say G is simply connected if $\mathbf{C}^* \setminus G$ is a connected subset of $\mathbf{C}^*$. (Many writers use this definition in the case when G is not assumed connected; connectedness of G is not in fact used until theorem B.4 below.)

If we look back at the example above, another condition, based on the topological index, becomes apparent. For if we require that for each closed path γ with $\gamma^* \subset G$, all points where $n(\gamma, w) \neq 0$ are in G, we remove the possibility which the example reveals. It is a surprising and unobvious fact that these two properties are equivalent.

Theorem B.2. *Let G be a connected open subset of* $\mathbf{C}$. *Then G is simply connected (that is,* $\mathbf{C}^* \setminus G$ *is connected) if, and only if, for each closed path γ, with $\gamma^* \subset G$, and each point $w \in \mathbf{C} \setminus G$, we have $n(\gamma, w) = 0$.*

Proof. We prove both implications by a contrapositive argument.

Suppose that there is a closed curve γ with $\gamma^* \subset G$, and a point

$$w \in \mathbf{C} \setminus G \quad \text{with} \quad n(\gamma, w) \neq 0.$$

We know from 5.5 that $n(\gamma, w) = 0$ if w is in the unbounded component of $\mathbf{C} \setminus \gamma^*$; hence w lies in a bounded component C_1 of $\mathbf{C} \setminus \gamma^*$. Let C_2 be the union of all the other components of $\mathbf{C} \setminus \gamma^*$ (including the unbounded component to which we will adjoin ∞). Then C_1, C_2 and γ^* are mutually disjoint and their union is $\mathbf{C}^*$. We have

$$w \in C_1 \cap (\mathbf{C}^* \setminus G), \quad \infty \in C_2 \cap (\mathbf{C}^* \setminus G), \quad \text{and} \quad \mathbf{C}^* \setminus G \subset C_1 \cup C_2,$$

since $\gamma^* \subset G$: it follows that $\mathbf{C}^* \setminus G$ is not connected.

Conversely, suppose that $\mathbf{C}^* \setminus G$ is not connected. We have to find a point $w \in \mathbf{C} \setminus G$ and a closed path in G with $n(\gamma, w) \neq 0$, and this argument, since it requires us to construct a path with certain properties, is necessarily complicated.

Since $\mathbf{C}^* \setminus G$ is not connected, there exist disjoint open sets G_1 and G_2 in $\mathbf{C}^*$, for which

$$E_1 = G_1 \cap (\mathbf{C}^* \setminus G) \quad \text{and} \quad E_2 = G_2 \cap (\mathbf{C}^* \setminus G)$$

are nonempty and

$$E_1 \cup E_2 = \mathbf{C}^* \setminus G.$$

Suppose for definiteness that $\infty \in E_1$, so that

$$E_2 = (\mathbf{C}^* \setminus G_1) \cap (\mathbf{C}^* \setminus G)$$

is a compact subset of $\mathbf{C}$, and G_2 is an open subset of $\mathbf{C}$ containing no points of $\mathbf{C}^* \setminus G$ except those in E_2. Hence it is sufficient, given a fixed point w of E_2, to construct a closed path in $G_2 \setminus E_2$ whose index with respect to w is nonzero.

Since E_2 is compact, and G_2 an open set containing E_2, it follows from 1.2 that there is a positive real number $\delta > 0$, such that if $z \in E_2$ and $z' \notin G_2$, then $|z' - z| \geqslant 2\delta$. It follows that if we superimpose a mesh of equally spaced lines in both coordinate directions whose separation is less than δ, then we divide the plane into small squares having the property that if any square intersects E_2, then it is contained in G_2. We can suppose that the mesh is located so that no line passes through w.

For each of the finite number $(Q_i)_{i=1}^n$, say, of squares which intersect E_2, let γ_i be the perimeter of the square, taken so that $n(\gamma_i, z) = 1$ if z is in the bounded component of $\mathbf{C} \setminus \gamma_i^*$ (compare 5.8). Let

$$\gamma = \gamma_1 \cup \gamma_2 \cup \cdots \cup \gamma_n,$$

and observe that every side which intersects E_2 is counted once in each direction, and so cancels out, and it follows that $\gamma^* \subset G_2 \setminus E_2$. An inductive argument on the number of squares shows that γ is the union of a number of disjoint closed paths, $\delta_1, \delta_2, \ldots, \delta_r$, say, and

$$\sum_{j=1}^{r} n(\delta_j, w) = n(\gamma, w) = \sum_{i=1}^{n} n(\gamma_i, w).$$

But the point w lies in exactly one of the squares Q_i, so that

$$\sum_{i=1}^{n} n(\gamma_i, w) = 1,$$

and at least one of the terms $n(\delta_j, w)$ is nonzero, which completes the proof.

The usefulness of this theorem is that it shows the equivalence of the 'weak' property of simple connectedness, that is the connectedness of $\mathbf{C}^* \setminus G$ which is easily recognized in examples, with the 'strong' property that no closed path in G can have nonzero index with respect to a point not in G. It is this strong property which we shall need in developing Cauchy's theorem to its fullest extent.

Some examples will illustrate the concept of simple connectedness.

Examples B.3. (i) The set $G = \mathbf{C} \setminus \bar{S}(0, 1) = \{z: |z| > 1\}$ is connected, but not simply connected. Notice that $\mathbf{C} \setminus G$ is connected, but

$$\mathbf{C}^* \setminus G = \{\infty\} \cup \bar{S}(0, 1)$$

is not.

(ii) The set $G = \{z: 0 < \operatorname{Re} z < 1\}$ illustrates another good reason for making definition B.1 in the way that we have. G is an infinite strip which one feels intuitively should be simply connected (it has no holes in it), but $\mathbf{C} \setminus G$ has two components. $\mathbf{C}^* \setminus G$ is connected, however.

(iii) The set $G = S(1, 1) \cup S(-1, 1)$ is not connected and so cannot be simply connected by our definition. However, it does have the property that $\mathbf{C}^* \setminus G$ is connected.

(iv) $G = \mathbf{C} \setminus C(0, 1)$ is not connected, nor is $\mathbf{C}^* \setminus G$.

We can now prove our main theorem which shows that simple connectedness is indeed the 'right' hypothesis for Cauchy's theorem.

Theorem B.4. *Let G be a connected open subset of* $\mathbf{C}$.

Then the following conditions on G are equivalent:

(i) *G is simply connected,*

(ii) *For each regular function f on G, and each closed path γ in G, we have*

$$\int_\gamma f(z)\,dz = 0,$$

(iii) *for each regular function f on G, there is a regular function F on G, with $F' = f$,*

(iv) *for each regular function f which has no zero in G, there is a regular function g for which $f = e^g$.*

Proof. The implication (i) $\Rightarrow$ (ii) is the full version of Cauchy's theorem and is much the most difficult of the assertions to prove. We begin with the others.

(ii) $\Rightarrow$ (iii) Let f be regular on G, and let z_0 be a fixed point of G. If z is any other point of G, then by 1.17, there is a path γ (in fact a polygonal path) in G with z_0 and z for initial and final points. If γ' is any other path in G from z_0 to z, then $\gamma \cup (-\gamma')$ is a closed path in G, and condition (ii) shows that

$$\int_{\gamma\cup(-\gamma')} f(z)\,dz = 0 \quad \text{or} \quad \int_\gamma f(z)\,dz = \int_{\gamma'} f(z)\,dz.$$

It follows that if we define $F(z) = \int_\gamma f(u)\,du$, then the definition depends on z (and z_0) but not on the path chosen to reach z from z_0. This shows that F is well defined, and the argument in theorem 2.3 proves that $F' = f$.

(iii) $\Rightarrow$ (iv) Let f be a regular function with no zero on G, and let z_0 be a fixed point of G. Then f'/f is regular on G, and so by (iii) there is a function F which is regular on G, with $F' = f'/f$. Since G is connected, F is unique to within an additive constant. The properties of the exponential function in 1.8, together with the fact that $f(z_0) \neq 0$ allow us to choose this constant in such a way that $\exp(F(z_0)) = f(z_0)$. Let $h(z) = \exp(F(z))$ for each $z \in G$.

Then

$$\left(\frac{f}{h}\right)' = \frac{f'h - h'f}{h^2} = \frac{f'\exp F - \exp F \cdot F'f}{h^2} = 0.$$

It follows that f/h is constant on G, and since $f(z_0) = h(z_0)$, the constant must be unity.

Hence for all $z \in G$, $f(z) = h(z) = \exp(F(z))$.

(iv) $\Rightarrow$ (i) Let γ be a closed path in G, and w a point of $\mathbf{C} \setminus G$.

The function f given by $f(z) = z - w$ for $z \in G$, is regular and nonzero on G. Condition (iv) shows that there is a function which is regular on G, with $f(z) = \exp(g(z))$ for all $z \in G$.

It follows that

$$g'(z) = \frac{f'(z)}{f(z)} = \frac{1}{z - w} \quad \text{for each } z \in G,$$

and hence by (v) of 1.20 that

$$n(\gamma, w) = \frac{1}{2\pi i}\int_\gamma \frac{dz}{z - w} = \frac{1}{2\pi i}\int_\gamma g'(z)\,dz = 0.$$

(i) $\Rightarrow$ (ii) This implication, like the proof of B.2 which it somewhat resembles, depends on the detailed examination of the way in which a path can be built up from a number of small rectangular components.

Suppose then that G is simply connected, that f is regular on G, and that γ is a closed path in G. Our first step is to replace γ by a simpler path, δ say. The compactness of γ^* as a subset of G implies the existence of a number $\eta > 0$, such that if $z \in \gamma^*$, then $S(z, \eta) \subset G$. Since γ is a uniformly continuous function from $[0, 1]$ to γ^*, we can find a finite number of points

$$0 = t_0 < t_1 < \cdots < t_{n-1} < t_n = 1$$

in $[0, 1]$ such that, if

$$t_{k-1} \leqslant t \leqslant t_k,\ k = 1, 2, \ldots, n,$$

then $|\gamma(t) - \gamma(t_{k-1})| < \eta$, or, if we let $z_k = \gamma(t_k)$, then

$$\gamma(t) \in S(z_{k-1}, \eta) \quad \text{if} \quad t_{k-1} \leqslant t \leqslant t_k.$$

Let δ_k be the path from z_{k-1} to z_k which consists of two sides of the rectangle whose sides are parallel to the coordinate axes and which has z_{k-1} and z_k for opposite vertices. We now have a closed curve, formed by following γ from z_{k-1} to z_k, and then $-\delta_k$ from z_k back to z_{k-1}, which lies in a starred domain (namely $S(z_{k-1}, \eta)$) on which f is regular.

Theorem 2.3 now shows that

$$\int_{t_{k-1}}^{t_k} f(\gamma(t))\gamma'(t)\,dt = \int_{\delta_k} f(z)\,dz.$$

It follows that if $\delta = \delta_1 \cup \delta_2 \cup \cdots \cup \delta_k$, then δ is a closed path in G, and on summing the above equation over k, that

$$\int_\gamma f(z)\,dz = \int_\delta f(z)\,dz.$$

The particular case when $f(z) = (z - w)^{-1}$ for $w \notin G$, shows that

$$n(\delta, w) = n(\gamma, w) = 0.$$

Our next step is to extend the line segments which make up δ in both directions, thereby forming a mesh of lines (which, unlike that used in B.2, need not be equally spaced) which decomposes **C** into a finite number of closed rectangular regions $R_1, R_2, \ldots, R_m$ say, as well as a number of unbounded regions. Since δ^* is bounded, δ is a union of certain edges of these rectangles, possibly repeated a number of times, in one or both directions.

For each $j = 1, 2, \ldots, m$, let a_j be the centre of R_j, and let $n_j = n(\delta, a_j)$ be the topological index of δ with respect to a_j. Let θ_j be the perimeter of R_j, taken in the sense which makes $n(\theta_j, a_j) = 1$ (compare 5.8). Clearly if $j \neq k$, $n(\theta_j, a_k) = 0$.

Now consider the path θ which consists of the union for all j of n_j iterates of θ_j (if $n_j > 0$, one transverses θ_j n_j times, if $n_j = 0$, θ_j is omitted, while if $n_j < 0$, one traverses $-\theta_j$ $|n_j|$ times). We shall show that

$$\int_\theta f(z)\,dz = \int_\delta f(z)\,dz.$$

For suppose that s is a common side of rectangles R_j and R_k, whose direction corresponds to that which occurs in θ_j (and so is opposite to that which occurs in θ_k). Suppose that in traversing δ, s is traversed p times in the positive sense, and q times in the negative sense. A $(p + q)$-fold application of theorem 5.7 shows that

$$n_j = n(\delta, a_j) = p - q + n(\delta, a_k) = (p - q) + n_k.$$

But in the expression of θ, an s occurs $(n_j - n_k)$ times $= (p - q)$ times, so that we have proved that a common side of two rectangles occurs the same number of times in both θ and δ. An analogous argument (which the reader is recommended to supply for himself) shows that a side which separates a rectangle R_j from an unbounded region also occurs the same number of times in both θ and δ. It follows that

$$\int_\delta f(z)\,dz = \int_\theta f(z)\,dz = \sum_{j=1}^{m} n_j \int_{\theta_j} f(z)\,dz.$$

We complete the proof by showing that each term on the right-hand side vanishes.

For suppose that for some j, $R_j \subset G$. Then we may enlarge R_j slightly to obtain a starred open set in G containing θ_j^*, and on which f is regular. Then theorem 2.3 shows that $\int_{\theta_j} f(z)\,dz = 0$. If on the other hand R_j is not contained in G, f may well not be defined on θ_j^*—however, we are about to prove that $n_j = 0$ in this case, so that no harm is done. Suppose that b_j is a point of R_j which is not in G. Since δ^* is a subset of G, it follows that the segment $[a_j, b_j]$ is disjoint from δ^*, and hence that

$$n(\delta, a_j) = n(\delta, b_j) = n(\gamma, b_j)$$

as was noted earlier. But our hypothesis of simple connectedness (together with B.2) shows that $n(\gamma, b_j) = 0$, and this completes the proof.

We conclude this section with a brief mention of another famous (and difficult) theorem which includes simple connectedness as one of its hypotheses. We discussed at the end of Chapter 5 the fact that a univalent regular function from G to $f(G)$ had an inverse which was also regular. Such a function establishes an equivalence relation amongst open sets in $\mathbf{C}$ by defining $G_1 \sim G_2$ if there is a univalent regular function f on G, with $G_2 = f(G_1)$. The surprising fact is that if we restrict our attentions to simply connected open sets (which are assumed connected by definition B.1), there are only two equivalence classes, one of which consists of $\mathbf{C}$ alone.

More precisely, if we take the unit disc $U = S(0, 1)$ as a typical simply connected set not equal to $\mathbf{C}$, then we have:

(A) There is no regular mapping of $\mathbf{C}$ onto U (this is an immediate corollary of Liouville's theorem), and

(B) The Riemann mapping theorem. *Let G be a simply connected open subset of $\mathbf{C}$ with $G \neq \mathbf{C}$. Then there is a univalent regular mapping of G onto U. For any $z_0 \in G$, we may satisfy the conditions $f(z_0) = 0$, $f'(z_0) > 0$ in which case the map is uniquely determined.*

Owing to the great variety of simply connected subsets of **C**, the proof of this result is highly non-constructive in character, and requires a notion of compactness applied to sets of functions, rather than sets of complex numbers. The interested reader is referred to the books mentioned in the bibliography for details.

§2. THE JORDAN CURVE THEOREM

We studied in Chapter 5, most particularly in relation to Rouché's theorem, the class of curves γ, for which $\mathbf{C} \setminus \gamma^*$ had a single bounded component. Many of these curves, circles for instance, or the perimeters of rectangles, had the property that they are non-self-intersecting: that is if t_1, t_2 are distinct points of $[0, 1)$, then $\gamma(t_1) \neq \gamma(t_2)$. A curve with this property is said to be *simple*. An equivalent property is that there should exist a continuous one to one function from $C(0, 1)$ to γ^*, or in other words that the two spaces are topologically indistinguishable, and this explains the importance which the concept has in the study of the topology of plane sets.

The condition that a curve should be simple is very easy to recognize, which probably explains its long popularity. On the other hand as soon as one attempts to prove any results using it as a hypothesis, one is immediately faced with severe difficulties which are of no relevance to analysis. The basic fact here is that the hypothesis of simplicity is sufficient (though not necessary, as some easy examples show) to guarantee that $\mathbf{C} \setminus \gamma^*$ has a single bounded component: in fact more is true and we state the result formally.

Jordan curve theorem. *Let γ be a simple (that is, non-self-intersecting) closed curve. Then $\mathbf{C} \setminus \gamma^*$ has exactly two components, one unbounded which will be denoted by E, and one bounded which we have called B: γ^* is the boundary of both E and B.*

Furthermore the value of $n(\gamma, w)$ for $w \in B$ is ± 1. (This requires either that γ is a path, or an extension of the idea of winding numbers to general closed curves.)

For a detailed proof which involves the use of complex integration, and winding numbers, the reader is referred to the appendix to chapter nine of the book by J. Dieudonné mentioned in the bibliography. The principal value of the theorem in complex analysis (if it is used at all) lies not so much in the information on the number of components, as in the evaluation of the winding number. It is precisely our criterion, contained in theorem 5.7, for the evaluation of winding numbers for a much wider class of closed paths which allows us to avoid having recourse to the Jordan curve theorem.

It is to be hoped that this famous theorem will become to be seen for what it really is: the cornerstone of the topology of plane sets, and not the bugbear of complex analysis.

Appendix C
Logarithms and Fractional Powers

This appendix is devoted to a treatment of the complex logarithmic function, which was not defined in the main text. The reason for the postponement is twofold. Firstly the fact that there are many ways of defining an 'inverse exponential function' means that there is a certain degree of arbitrariness involved in settling on any one of them, and secondly with a little ingenuity, one can avoid mentioning logarithms until after the properties of winding numbers are established, thereby hopefully avoiding the temptation to 'evaluate' the integral

$$\frac{1}{2\pi i}\int_{\gamma}\frac{dz}{z-w}$$

by a logarithmic substitution.

We begin by rapidly reviewing those properties of the real logarithmic function which we wish to generalize. For real x, the exponential function $g: x \to e^x$ is strictly increasing, strictly positive, and has the properties $g(0) = 1$, $g' = g$. It follows that the inverse function f of g is a well-defined function from $\mathbf{R}^+ = \{x: x > 0\}$ to $\mathbf{R}$, which satisfies $f(1) = 0$, and by the rule for differentiating composite functions, if $f(x) = y$ $(x > 0)$, then $f(g(y)) = y$ and so

$$1 = (f \circ g)'(y) = f'(g(y)) \cdot g'(y) = f'(x)g(y) = xf'(x).$$

Hence for $x > 0$, $f'(x) = 1/x$, and $f(1) = 0$. It follows that if $x > 0$, another expression for f is given by

$$f(x) = \int_1^x \frac{dt}{t},$$

and from this (or directly from the corresponding properties of the exponential function) we can deduce that

(a) for $x, y > 0$, $f(xy) = f(x) + f(y)$;
(b) $f(x) \to +\infty$ as $x \to +\infty$, $f(x) \to -\infty$ as $x \to 0$;
(c) if $k > 0$, $x^{-k}f(x) \to 0$ as $x \to +\infty$, $x^k f(x) \to 0$ as $x \to 0$.

It turns out that it is not possible to define a function on **C** which has all these properties. On **C**, the exponential function is no longer one-to-one:

$$\exp z = \exp(z + 2k\pi i) \quad \text{for all} \quad z \in \mathbf{C}, \quad \text{and integers } k,$$

and so an inverse function cannot be defined without making a choice between several points at each of which the exponential function has the same value. Another possibility is to define $f(z) = \int_\gamma dw/w$, where γ is a path from 1 to z which obviously must avoid the origin. However, there are many ways to choose the path, and the fact that $\int_{C(0,1)} dw/w = 2\pi i$ shows that they will not all give the same answer. We avoid this difficulty by making the following definition.

Definition C.1. (The logarithmic functions). Let G be a starred open set in **C**, which contains $\mathbf{R}^+$, but not 0. The proof of theorem 2.3 shows that there is a regular function F on G for which $F'(z) = 1/z$. This function, which is uniquely determined if we require that $F(1) = 0$, will be denoted by $\log_G(z)$.

Notice that although different choices of G give rise to different functions $\log_G$, the assumption that $G \supset \mathbf{R}^+$ together with $F(1) = 0$ ensure that for x real and positive, $\log_G(x) = \log x$.

[It is a slightly tricky procedure to deduce from this definition that if $w = \log_G z$, then $\exp w = z$. One argument is as follows.

Let γ be a path in G with $\gamma(0) = 1$, $\gamma(1) = z$. Write

$$|\gamma(t)| = \rho(t) > 0 \quad \text{and} \quad \gamma(t) = \rho(t)\, e^{i\theta(t)},$$

where θ is a real continuous function on [0, 1]—it is the continuity of θ which causes difficulty. Then ρ and θ are piecewise continuously differentiable and we have

$$w = \log_G z = \int_\gamma \frac{du}{u} = \int_0^1 \frac{\gamma'(t)\,dt}{\gamma(t)} = \int_0^1 \frac{\rho'(t)\,e^{i\theta(t)} + i\rho(t)\,e^{i\theta(t)}\theta'(t)}{\rho(t)\,e^{i\theta(t)}}\,dt$$

$$= \int_0^1 \frac{\rho'(t)}{\rho(t)}\,dt + i\int_0^1 \theta'(t)\,dt = \log \rho(1) + i\theta(1).$$

Hence $\exp w = \rho(1)\,e^{i\theta(1)} = z$.]

A more general approach would be to use the results of Appendix B and let G be any simply connected open set not equal to $\mathbf{C}$, w be a point of $\mathbf{C} \setminus G$, and z_0 a fixed point of G. Then there is a unique regular function F (which we might denote by $\log_{G,z_0}(z - w)$) with $F'(z) = (z - w)^{-1}$ on G, and $F(z_0) = 0$. We shall not investigate this possibility further.

The known values for the derivatives of the real logarithmic function at $x = 1$ show that the power series expansion of $\log_G$ about $z = 1$ is the familiar

$$\sum_{n=1}^{\infty} (-1)^{n-1} \frac{(z-1)^n}{n} \quad \text{which converges on } S(1, 1).$$

Taylor's theorem shows that the sum is $\log_G z$ provided $S(1, 1) \subset G$.

Exercise C.2. Find the power series expansion about any other point of G.

What is its radius of convergence? Compare with example 3.3 (ii) which deals with the function we could now write as $\log_D (1 + z)$.

Normally one chooses for G a set of the form

$$G_\theta = \mathbf{C} \setminus \{z : z = \lambda\,e^{i\theta}, \quad 0 < \lambda < \infty\}$$

for some θ, $0 < \theta < 2\pi$, and writes $\log_\theta z$ instead of the more cumbersome $\log_{G_\theta} z$. In particular, the choice $\theta = \pi$ defines the principal logarithm which will simply be denoted $\log z$ without any subscript. The properties of the exponential function, listed under 1.8 show that any complex number z which is not real and negative may be written in a unique way as $z = \rho\,e^{i\alpha}$, where $\rho > 0$, and $-\pi < \alpha < \pi$. We now identify $\log z = \log_\pi z$ in terms of ρ and α. Let γ be the path from 1 to z which comprises the segment $[1, \rho]$ of the real axis, followed by the arc of $C(0, \rho)$ from ρ to z which lies in G_π. Then

$$\log z = \int_\gamma \frac{dw}{w} = \int_1^\rho \frac{dx}{x} + \int_0^\alpha \frac{i\rho\,e^{i\beta}}{\rho\,e^{i\beta}}\,d\beta = \log \rho + i\alpha,$$

(where naturally $\log \rho$ refers to the real logarithmic function) which verifies that $\exp(\log z) = \rho\, e^{i\alpha} = z$ in this particular case. This could have been used as the defining property for $\log_\pi$, and then the other functions $\log_\theta$ could have been determined in terms of $\log_\pi$ (exercise—work out the definition of $\log_\theta$ in terms of $\log_\pi$). However, the definition using Cauchy's theorem to find a primitive for the function $(z \to z^{-1})$ seems more satisfying and explains how the many possible choices for a logarithmic function arise. Notice that although we have arranged matters so that

$$\exp(\log z) = z \quad \text{for all} \quad z \in G_\pi,$$

we have $\log(e^w) = w$ if and only if $|\mathrm{Im}\, w| < \pi$.

We now investigate to what extent property (a) continues to hold for the principal logarithm.

Theorem C.3. *Let* log *denote the principal logarithm, defined on* G_π. *If* $\mathrm{Re}\, z > 0$, $\mathrm{Re}\, w > 0$, *then* $zw \in G_\pi$, *and*

$$\log(zw) = \log z + \log w.$$

Proof. Suppose $z = x + iy$, $w = u + iv$, where $x, u > 0$.

If $\mathrm{Im}(zw) = xv + uy \neq 0$ then $zw \in G_\pi$.

If $\mathrm{Im}(zw) = 0$, $v = -(u/x)y$, and so

$$\mathrm{Re}(zw) = xu - yv = \frac{u}{x}(x^2 + y^2) > 0$$

and so $z \in G_\pi$ again.

Let $z = \rho_1 e^{i\alpha_1}$, $\rho_1 > 0$, $|\alpha_1| < \pi$. Then since $\mathrm{Re}\, z = \rho_1 \cos \alpha_1 > 0$, we must have $|\alpha_1| < \frac{1}{2}\pi$. Similarly

$$w = \rho_2 e^{i\alpha_2}, \quad \rho_2 > 0, \quad |\alpha_2| < \tfrac{1}{2}\pi, \quad \text{and} \quad zw = \rho_1\rho_2\, e^{i(\alpha_1 + \alpha_2)}.$$

If follows that $\rho_1\rho_2 > 0$, and $|\alpha_1 + \alpha_2| < \pi$ (which provides us with an alternative proof that $zw \in G_\pi$).

Then

$$\begin{aligned} \log(zw) &= \log \rho_1\rho_2 + i(\alpha_1 + \alpha_2) \\ &= (\log \rho_1 + i\alpha_1) + (\log \rho_2 + i\alpha_2) = \log z + \log w. \end{aligned}$$

It is easy to see that serious trouble ensues if the hypothesis $\mathrm{Re}\, z$ and $\mathrm{Re}\, w > 0$ is dropped. For instance, $i \in G_\pi$, while $i^2 = -1$ does not. Furthermore

$$\omega = e^{i2\pi/3} \in G_\pi \quad \text{and} \quad \log \omega = 2\pi i/3,$$

but

$$\omega^2 \in G_\pi, \qquad \log(\omega^2) = -i\frac{2\pi}{3} \neq i\frac{4\pi}{3}.$$

Properties (b) and (c) remain valid for complex logarithms: the precise statement and verification are left to the reader.

Definition C.4. (The argument functions). Let $\theta \in (0, 2\pi)$, and $z \in G_\theta$. Then the argument function, $\arg_\theta$ is defined by

$$\arg_\theta z = \operatorname{Im}(\log_\theta z).$$

The particular function $\arg_\pi$ is called the principal argument, and is denoted simply by arg.

A consequence of theorem 6.3 is that

$$\arg(zw) = \arg z + \arg w \quad \text{if} \quad \operatorname{Re} z > 0, \quad \operatorname{Re} w > 0.$$

Notice that for any values of θ, ϕ, $\log_\theta z$ and $\log_\phi z$ (if both defined) can differ by at most $2\pi i$, and so $\arg_\theta z$ and $\arg_\phi z$ can differ by at most 2π.

The definitions of the logarithmic and argument functions allow us to define complex powers of complex numbers in such a way that some of the corresponding properties for integer powers are carried over.

Notice that if n is an integer,

$$z^n = \{\exp(\log_\theta z)\}^n = \exp(n \log_\theta z)$$

by 1.8(a). Since $\log_\theta z$ is unique to within $2\pi i$, this expression is independent of θ. For non-integer powers, the situation is less fortunate.

Definition C.5. (Complex powers). Let z, w be complex numbers, with $z \in G_\pi$. Define the power z^w by

$$z^w = \exp(w \log z).$$

In fact what we have defined here is the principal value of the power—other values could be obtained from the expression

$$\exp(w \log_\theta z),$$

defined on G_θ. [Some such expedient is inevitable if one wishes to assign a meaning to a non-integer power of a negative real number. For instance one determination of $(-1)^{1/2}$ would be

$$\exp\left(\tfrac{1}{2}\log_{-\pi/2}(-1)\right) = \exp\left(\tfrac{1}{2}i\pi\right) = i.]$$

The main properties of the complex powers are listed in the following.

Theorem C.6

(i) *For* $z \in G_\pi$, *and* $w_1, w_2 \in \mathbf{C}$,

$$z^{w_1+w_2} = z^{w_1}z^{w_2};$$

(ii) *For* $\operatorname{Re} z_1 > 0$, $\operatorname{Re} z_2 > 0$, *and* $w \in \mathbf{C}$,

$$(z_1z_2)^w = z_1^w \cdot z_2^w;$$

(iii) For $z \in G_\pi$, $w_1, w_2 \in \mathbf{C}$ and $|\operatorname{Im}(w_1 \log z)| < \pi$,

$$(z^{w_1})^{w_2} = z^{w_1w_2};$$

(iv) *For fixed* $w \in \mathbf{C}$, *let* $f(z) = z^w$ *for* $z \in G_\pi$. *Then* f *is regular on* G_π, *and* $f'(z) = wz^{w-1}$ *for all* $z \in G_\pi$.

(v) *For fixed* $z \in G_\pi$, *let* $g(w) = z^w$ *for* $w \in \mathbf{C}$. *Then* g *is regular on* $\mathbf{C}$ *and* $g'(w) = \log z \cdot z^w$ *for all* $w \in \mathbf{C}$.

Proof. (i) is an immediate consequence of definition C.5, and (ii) of C.5 and C.3.

(iii) The condition $|\operatorname{Im}(w_1 \log z)| < \pi$ ensures that $z^{w_1} \in G_\pi$, and so $(z^{w_1})^{w_2}$ is defined. The observation made preceding C.3 now shows that

$$\log(\exp(w_1 \log z)) = w_1 \log z$$

and hence that

$$\begin{aligned}(z^{w_1})^{w_2} = \exp\{w_2 \log(z^{w_1})\} &= \exp\{w_2 \log(\exp(w_1 \log z))\} \\ &= \exp(w_2w_1 \log z) = z^{w_1w_2}.\end{aligned}$$

(iv) For $z \in G_\pi$, $w \in \mathbf{C}$,

$$f(z) = \exp(w \log z),$$

and since log is regular on G_π (its derivative, by definition C.1 is $z \to z^{-1}$), we deduce that f is regular. Also

$$f'(z) = \exp(w \log z) \cdot wz^{-1} = wz^w \cdot z^{-1} = wz^{w-1} \quad \text{by (i).}$$

(v) For $z \in G_\pi$, $w \in \mathbf{C}$, let

$$g(w) = z^w = \exp(w \log z).$$

Then z (and so $\log z$) are constant, so that g is regular as a function of w, and

$$g'(w) = \exp(w \log z) \cdot \log z = \log z \cdot z^w.$$

If we wish to define logarithms or complex powers without the rather messy restrictions which we have placed on them, then it becomes necessary to 'stick together' different pieces of the function in such a way that it retains its regularity and single valued-ness, but is defined on a set which is more general than a subset of the complex plane. For this, the reader is referred to the monograph *The Concept of a Riemann Surface* by H. Weyl, published by Addison Wesley, or to the more advanced books mentioned in the bibliography.

Bibliography

For the prerequisites in elementary analysis, two books which are both thorough and readable are

> *Calculus*, by M. Spivak, published by W. Benjamin Inc., New York,

and

> *The Principles of Mathematical Analysis*, by W. Rudin, published by McGraw-Hill.

The first of these is mainly concerned with analysis of functions of a single real variable, while the second treats analysis in a more abstract metric space setting. Further metric space theory and its application to the theory of general linear spaces may be found in

> *The Elements of Functional Analysis*, by I. J. Maddox, published by Cambridge University Press.

Undoubtedly the most influential and authoritative text on complex analysis itself is

> *Complex Analysis* (Second Edition), by L. V. Ahlfors, published by McGraw-Hill.

A student with sufficient mathematical maturity to read this will have no need of an introductory text such as the present. Another very lucid treatment of the subject is contained in

> *Real and Complex Analysis*, by W. Rudin, published by McGraw-Hill.

In particular this contains proofs of important theorems on the approximation of regular functions by polynomials and rational functions.

Innumerable problems which are both instructive and enjoyable are in

> *Aufgaben and Lehrsätze aus der Analysis*, by G. Polya and G. Szegö, published by J. Springer, Berlin, and recently translated into English.

Indeed, the whole of this two-volume work, which includes solutions to all the problems, could be used as an heroic course in 'teach yourself analysis'.

For a treatment of more advanced topics, particularly in the field of functions which are regular in the whole complex plane, there is

> *Analytic Functions*, by R. Nevanlinna, published by J. Springer, Berlin.

Finally, a comprehensive synthesis of many of the diverse topics which nowadays come under the heading of analysis is expounded in

> *The Foundations of Modern Analysis*, by J. Dieudonné, published by Academic Press, New York,

of which the first three volumes are now available.

Index

Index of notations